Drought Assessment, Management, and Planning

NATURAL RESOURCE MANAGEMENT AND POLICY

Editors:

Ariel Dinar
Dept. of Agricultural Economics
University of California, Davis
Davis, CA 95616-8512

David Zilberman
Dept. of Agricultural and
Resource Economics
University of California, Berkeley
107 Giannini Hall
Berkeley, CA 94720

EDITORIAL STATEMENT

There is a growing awareness to the role that natural resources such as water, land, forests and environmental amenities play in our lives. There are many competing uses for natural resources, and society is challenged to manage them for improving social well being . Furthermore, there may be dire consequences to natural resources mismanagement. Renewable resources such as water, land and the environment are linked, and decisions made with regard to one may affect the others. Policy and management of natural resources now require interdisciplinary approach including natural and social sciences to correctly address our society preferences.

This series provides a collection of works containing most recent findings on economics, management and policy of renewable biological resources such as water, land, crop protection, sustainable agriculture, technology, and environmental health. It incorporates modern thinking and techniques of economics and management. Books in this series will incorporate knowledge and models of natural phenomena with economics and managerial decision frameworks to assess alternative options for managing natural resources and environment.

This book is a timely collection of theory and case studies addressing drought assessment, management and planning. The book includes up-to-date techniques used to cope with drought at local, regional and national levels. The book includes also case studies from countries which have a long-standing history of drought, and have accumulated drought experience over time. This experience is shared with the reader for a better understanding of future drought awareness.

The Series Editors

Previously Published Books in the Series

Russell, Clifford S. and Shogren, Jason F.:
Theory, Modeling and Experience in the Management of Nonpoint-Source Pollution

Drought Assessment, Management, and Planning:

Theory and Case Studies

edited by
Donald A. Wilhite

KLUWER ACADEMIC PUBLISHERS
BOSTON / DORDRECHT / LONDON

Distributors for North America:
Kluwer Academic Publishers
101 Philip Drive
Assinippi Park
Norwell, Massachusetts 02061 USA

Distributors for all other countries:
Kluwer Academic Publishers Group
Distribution Centre
Post Office Box 322
3300 AH Dordrecht, THE NETHERLANDS

Library of Congress Cataloging-in-Publication Data

Drought assessment, management, and planning : theory and case studies / edited by Donald A. Wilhite.
p. cm.
Includes index.
ISBN 0-7923-9337-6
1. Droughts. 2. Droughts--Case studies. 3. Droughts--Management. 4. Droughts--Government policy. I. Wilhite, Donald A.
QC929.24.D76 1993
363.3'492--dc20 93-22111
CIP

Printed on acid-free paper.

Printed in the United States of America

To my wife, Myra

Contents

Part Two: Case Studies

Preface

Drought is an insidious hazard of nature. It originates from a deficiency of precipitation that results in a water shortage for some activity or some group. Africa has suffered the most dramatic impacts from drought during the past several decades—the recent droughts in the southern and eastern portions of the continent are testimony to that fact. However, the vulnerability of all nations to extended periods of water shortage has been underscored again and again during this same time period. In the past decade alone, droughts have occurred with considerable frequency and severity in most of the developed and developing world. Significant parts of North and South America, Australia, Europe, and Asia have been plagued recently by extended periods of severe drought, often resulting in far-reaching economic, social, and environmental consequences. In the western United States, for example, vast areas are facing the prospects of a sixth or seventh consecutive year of drought in 1993.

Concern by members of the scientific and policy communities about the inability of governments to respond in an effective and timely manner to drought and its associated impacts exists worldwide. Numerous "calls for action" for improved drought planning and management have been issued by national governments, professional organizations, intergovernmental organizations, nongovernmental organizations, and others. The United Nations' International Decade for Natural Disaster Reduction (IDNDR) is yet another example of an international call for action to reduce the impacts that result from drought and other natural hazards. However, the task of altering the perception of policy makers and scientists about drought and educating them about alternative management approaches is a formidable challenge. Governments typically treat drought as a rare and random event that is inherently unpredictable; they are unprepared to respond effectively when it occurs. Effective drought response requires long-range planning, a difficult assignment for most governments.

Many scientists and policy makers now have an improved understanding of drought and its economic, social, and environmental impacts. Although impediments to drought planning exist, recent progress has created a sense of cautious optimism that society is steadily moving toward a higher level of preparedness. Also, technologies and information are available that would enable countries to more effectively mitigate the effects of drought through the development of a more proactive and systematic risk management approach.[1]

[1] For example, see D. A. Wilhite and W. E. Easterling, eds. 1987. *Planning for Drought: Toward a Reduction of Societal Vulnerability*. Westview Press, Boulder, Colorado, U.S.A.

Awareness, understanding, and appreciation of drought and its effects within the scientific and policy communities has been enhanced in the past decade through a series of notable events. First, the occurrence of severe drought worldwide during and following the extreme El Niño/Southern Oscillation (ENSO) event of 1982-83 captured the attention of governments by highlighting the vulnerability of all nations. The drought of 1988 in the United States, for example, resulted in costs and losses of nearly $40 billion.[2] Second, it appears that societal vulnerability to drought is increasing, largely because of population growth and society's increasing demand and competition for limited water and other natural resources. Population growth is putting added pressure on fragile lands. Third, many governments now have a better appreciation of the costs associated with drought. These costs include not only the direct impacts of drought but also the indirect costs (i.e., personal hardship, relief costs, retardation of economic development, and accelerated environmental degradation). Nations can no longer afford to allocate scarce financial resources to often short-sighted response programs that do little to mitigate, and may in fact exacerbate, the effects of future drought.

Given society's increasing vulnerability to drought and the uncertainties about future changes in climate, including the possible increased frequency and severity of extreme climatic events, improving the level of preparedness for the recurrence of drought represents an important step forward. Walter Orr Roberts, president emeritus of the National Center for Atmospheric Research in Boulder, Colorado, recently stated, "One logical approach, even in the face of great uncertainty about future climate, is to improve our ability to manage current climate extremes. Unlike global control of greenhouse gas emissions, international cooperation is not needed for many such adjustments to succeed; thus nations or regions that instigate them will be rewarded unilaterally."[3]

This book is dedicated to furthering scientific understanding of drought assessment and management and the policy issues associated with drought preparedness worldwide. The authors invited to contribute to this volume are uniquely qualified to address their respective topics. The book is divided into two parts. Part One emphasizes theoretical aspects of drought assessment, management, and preparedness; it also includes a discussion of the concept of drought. These chapters are aimed at improving

[2] A case study of the 1987-89 drought in the United States was completed recently (1990) by W. E. Riebsame, S. A. Changnon, and T. R. Karl, *Drought and Natural Resources Management in the United States* (Westview Press, Boulder, Colorado, U.S.A.).

[3] Roberts, W. O. 1990. Global climate change as a hazard [editorial]. *National Hazards Observer* 13(6):1-2. Natural Hazards Research and Applications Information Center, University of Colorado, Boulder, Colorado, U.S.A.

the reader's understanding and awareness of drought; appropriate technologies to monitor drought's onset, development, and termination; emerging methodologies to estimate impacts; adjustment strategies to alleviate drought effects; and a methodology for countries to follow in the development of a national drought policy and plan. Part One concludes with an insightful examination of how changes in climate may alter climatic variability and the frequency and intensity of extreme events, particularly drought.

The successes and failures of previous assessment and response efforts are highlighted in Part Two through case studies of Israel, India, Brazil, South Africa, Australia, the United States, and China. Several of these case studies provide insights into the difficulties inherent in the provision of timely and effective drought assistance by government in both a developed and developing country setting. The lessons learned in these instances provide both an extraordinary opportunity and challenge to other drought-prone nations to learn from these experiences and avoid some of the mistakes of the past. Some of the case studies in Part Two illustrate remarkable progress in drought preparedness and are examples for other nations to follow.

If society is to diminish vulnerability to future episodes of drought, then we must be better prepared for its inevitable recurrence and implement policies and plans that promote the minimization of risk in agricultural and other weather-sensitive industries. It is only through this effort that society will be able to reduce vulnerability to drought.

Donald A. Wilhite

Acknowledgments

The task of editing a book is a formidable one. In this case, the task was made much easier as a result of the cooperation of those authors selected to contribute to this volume. All of the authors enthusiastically accepted my invitation to participate in this project and completed their assigned chapters on schedule. I gratefully acknowledge their time and effort.

I am also indebted to my very capable editorial assistant, Deborah A. Wood. She is to be commended for her outstanding editorial skills and ability to manage and complete complex and difficult projects. Deborah has worked with me for more than ten years and her outstanding contributions to my research program are appreciated sincerely.

Finally, I wish to express my gratitude to my department chair and friend, Blaine L. Blad, for his support and encouragement of all that I seek to accomplish in my professional career.

Drought Assessment, Management, and Planning

Part One

Theory

Chapter 1
The Enigma of Drought

Donald A. Wilhite

INTRODUCTION

Drought is the most complex and least understood of all natural hazards, affecting more people than any other hazard (Hagman, 1984). For the past several decades, we have been reminded again and again of the ravages of drought and the inability of most societies to effectively mitigate impacts in the short term and reduce vulnerability in the longer term. In fact, most scientists would agree that vulnerability to drought is increasing for a number of reasons, the most important of which may be the increasing pressure of an expanding population base on limited water and other natural resources.

The purpose of this chapter is to lay the foundation for an understanding of the concept of drought. The primary emphasis of the chapter will be on understanding the concept of drought and why, according to Hagman (1984), the phenomenon is not better understood by scientists and policy makers. Drought is a normal part of climate and its recurrence, like other extreme climatic events, is inevitable. Through a better understanding and awareness of the characteristics of drought and its differences from other natural hazards, both scientists and policy makers will be better equipped to establish much-needed policies and plans whereby vulnerability can be reduced or stabilized for future generations.

DROUGHT: AN OVERVIEW

Drought differs from other natural hazards (e.g., floods, tropical cyclones, and earthquakes) in several ways. First, since the effects of drought often accumulate slowly over a considerable period of time and may linger for years after the termination of the event, a drought's onset and end are difficult to determine. Because of this, drought is often referred to as a "creeping phenomenon" (Tannehill, 1947). Second, the absence of a precise and universally accepted definition of drought adds to the confusion about whether or not a drought exists and, if it does, its degree of severity. Realistically, definitions of drought must be region and application (or impact) specific. This is one explanation for the scores of definitions that have been developed. Unfortunately, many of these definitions have not adequately defined drought in meaningful terms for scientists and policy makers. This is the result, at least in part, of misunderstandings of the concept by those formulating definitions. Third, drought impacts are less obvious and are spread over a larger geographical area than are damages that result from other natural hazards. Drought seldom results in structural damage. For these reasons, the quantification of impacts and the provision of disaster relief are far more difficult tasks for drought than they are for other natural hazards. These characteristics have hindered the development of accurate, reliable, and timely estimates of drought severity and impacts and, ultimately, the formulation of drought contingency plans by most governments.

Drought is a normal part of climate for virtually all climatic regimes. It occurs in high as well as low rainfall areas. Drought differs from aridity in that the latter is restricted to low rainfall regions and is a permanent feature of the climate. Many people associate the occurrence of drought with the Great Plains of North America, east Africa, west African Sahel, India, or Australia; they may have difficulty visualizing drought in Southeast Asia, Brazil, western Europe, or the eastern United States, regions perceived by many to have a surplus of water. For example, residents of many humid regions often refer to "green droughts" (i.e., droughts associated with apparent ample rainfall but reduced agricultural productivity because of poor timing of rains or ineffective precipitation). Thus, the character of drought is distinctly regional, reflecting unique meteorological, hydrological, agricultural, and socioeconomic characteristics.

Drought is the consequence of a natural reduction in the amount of precipitation received over an extended period of time, usually a season or more in length, although other climatic factors (such as high temperatures, high winds, and low relative humidity) are often associated with it in many regions of the world and can significantly aggravate the severity of the event. Drought is also related to the timing (i.e., principal season of occurrence, delays in the start of the rainy season, occurrence of rains in relation to principal crop growth stages) and the effectiveness of the rains (i.e., rainfall intensity, number of rainfall events).

Drought severity is dependent not only on the duration, intensity, and geographical extent of a specific drought episode, but also on the demands made by human activities and vegetation on a region's water supplies. The characteristics of drought, along with its far-reaching impacts, make its effects on society, economy, and environment difficult, though not impossible, to identify and quantify. The significance of drought should not be divorced from its societal context. The impact of a drought depends largely on societal vulnerability at that particular moment. Subsequent droughts in the same region will have different effects, even if they are identical in intensity, duration, and spatial characteristics.

DROUGHT TYPES AND DEFINITIONS

Because drought affects so many economic and social sectors, scores of definitions have been developed by a variety of disciplines. In addition, because drought occurs with varying frequency in nearly all regions of the globe, in all types of economic systems, and in developed and developing countries alike, the approaches taken to define it also reflect regional differences as well as ideological differences. Impacts also differ spatially and temporally, depending on the societal context of drought. A universal definition of drought is an unrealistic expectation.

Definitions of drought can be categorized broadly as either conceptual or operational (Wilhite and Glantz, 1985). Conceptual definitions are of the "dictionary" type, generally defining the boundaries of the concept of drought, and thus are generic in their description of the phenomenon. For example, the American Heritage Dictionary (1976) defines drought as "a long period with no rain, especially during a planting season." Operational definitions attempt to identify the onset, severity, continuation, and termination of drought episodes. Definitions of this type are often used in an "operational" mode. These definitions can also be used to analyze drought frequency, severity, and duration for a given historical period. An operational definition of agricultural drought might be one that compares daily precipitation to evapotranspiration (ET) rates to determine the rate of soil water depletion and then expresses these relationships in terms of drought effects on plant behavior at various stages of development. The effects of these meteorological conditions on plant growth would be reevaluated continuously by agricultural specialists as the growing season progresses.

Many disciplinary perspectives of drought exist. Each discipline incorporates different physical, biological, and/or socioeconomic factors in its definition of drought. Because of these numerous and diverse disciplinary views, considerable confusion often exists over exactly what constitutes a drought (Glantz and Katz, 1977). Research has shown that the lack of a precise and objective definition in specific situations has been an obstacle to understanding drought, which has led to indecision and/or inaction on the part of managers, policy makers, and others (Wilhite et al., 1986). It must be accepted that the importance of drought lies in its impacts. Thus definitions should be region and impact or application specific in order to be used in an operational mode by decision makers. A comprehensive review of drought definitions and indices can be found in a technical note published by the World Meteorological Organization (WMO) (1975). Consult Subrahmanyam (1967), Glantz and Katz (1977), Sandford (1979), Dracup et al. (1980), and Wilhite and Glantz (1985) for a thorough discussion of the difficulties in defining drought.

Drought can be grouped by type as follows: meteorological, hydrological, agricultural, and socioeconomic (Wilhite and Glantz, 1985). Meteorological drought is expressed solely on the basis of the degree of dryness (often in comparison to some "normal" or average amount) and the duration of the dry period. Definitions of meteorological drought must be considered as region specific since the atmospheric conditions that result in deficiencies of precipitation are highly variable from region to region. For example, some definitions differentiate meteorological drought on the basis of the number of days with precipitation less than some specified threshold. Such a definition is unrealistic in those regions in which extended periods without rainfall are common. Other definitions may relate actual precipitation departures to average amounts on monthly, seasonal, water year, or annual time scales. Definitions derived for application to one region usually are not transferrable to another since meteorological characteristics differ. Human perceptions of these conditions are equally variable. Both of these points must be taken into account in order to identify the characteristics of drought and make comparisons between regions.

Hydrological droughts are related more to the effects of periods of precipitation shortfall on surface or subsurface water supply (i.e., streamflow, reservoir and lake levels, groundwater) than to precipitation shortfalls (Dracup et al., 1980; Klemes, 1987). Hydrological droughts are usually out of phase or lag the occurrence of meteorological and agricultural droughts. Meteorological droughts result from precipitation deficiencies; agricultural droughts are largely the result of soil moisture deficiencies. More time elapses before precipitation deficiencies show up in components of the hydrological system (e.g., reservoirs, groundwater). As a result, impacts are out of phase with those in other economic sectors. Also, water in hydrological storage systems (e.g., reservoirs, rivers) is often used for multiple and competing purposes (e.g., power generation, flood control, irrigation, recreation), further complicating the sequence and quantification of impacts. Competition for water in these storage systems escalates during drought, and conflicts between water users increase significantly.

The frequency and severity of hydrological drought is often defined on the basis of its influence on river basins. Whipple (1966) defined a drought year as one in which the aggregate runoff is less than the long-term average runoff. Low-flow frequencies have been determined for many streams. If the actual flow for a selected time period falls below a certain threshold, then hydrological drought is considered to be in progress. However, the number of days and the level of probability that must be exceeded to define a hydrological drought period is somewhat arbitrary. These criteria will vary between streams and river basins.

Agricultural drought links various characteristics of meteorological and hydrological drought to agricultural impacts, focusing on precipitation shortages, differences between actual and potential evapotranspiration, soil water deficits, and so forth. A plant's demand for water is dependent on prevailing weather conditions, biological characteristics of the specific plant, its stage of growth, and the physical and biological properties of the soil. An operational definition of agricultural drought should account for the variable susceptibility of crops at different stages of crop development. For example, deficient subsoil moisture in an early growth stage will have little impact on final crop yield if topsoil moisture is sufficient to meet early growth requirements. However, if the deficiency of subsoil moisture continues, a substantial yield loss may result.

Finally, socioeconomic drought associates the supply and demand of some economic good or service with elements of meteorological, hydrological, and agricultural drought. Some scientists suggest that the time and space processes of supply and demand are the two basic processes that should be included in an objective definition of drought (Yevjevich, 1967). For example, the supply of some economic good (e.g., water, hay, hydroelectric power) is weather dependent. In most instances, the demand for that good is increasing as a result of increasing population and/or per capita consumption. Therefore, drought could be defined as occurring when the demand exceeds supply as a result of a weather-related supply shortfall (Sandford, 1979). This concept of drought supports the strong symbiosis that exists between drought and human activities. Thus, the incidence of drought could increase because of a change in the frequency of the physical event, a change in societal vulnerability to water shortages, or both. For example, poor land-use practices such as overgrazing can decrease animal carrying capacity and increase soil erosion, which exacerbates the impacts of and vulnerability to future droughts. This example is especially relevant in semiarid regions (e.g., Australia) and in areas of hilly or sloping terrain (e.g., Lesotho).

DROUGHT CHARACTERISTICS AND SEVERITY

Droughts differ from one another in three essential characteristics—intensity, duration, and spatial coverage. Intensity refers to the degree of the precipitation shortfall and/or the severity of impacts associated with the shortfall. It is generally measured by the departure of some climatic index from normal and is closely linked to duration in the determination of impact. The simplest index in widespread use is the percent of normal precipitation. With this index, actual precipitation is compared to "normal" or average precipitation for time periods ranging from 1 to 12 or more months. Actual precipitation departures are normally compared to expected or average amounts on a monthly, seasonal, annual, or water year (October-September) time period. One of the principal difficulties with this (or any) index is the choice of the threshold below which the deficiency of precipitation must fall (e.g., 75% of normal) to define the onset of drought. Thresholds are usually chosen arbitrarily. In reality, they should be linked to impact. Many indices of drought are in widespread use today, such as the decile approach (Gibbs, 1967; Lee, 1979; Coughlan, 1987) used in Australia, the Palmer Drought Severity Index and Crop Moisture Index (Palmer, 1965 and 1968) in the United States, and the Yield Moisture Index (Jose et al., 1991) in the Philippines and elsewhere. For a comparison of several popular meteorological indices, see Olidapo (1985).

Another distinguishing feature of drought is its duration. Droughts usually require a minimum of 2-3 months to become established but then can continue for several consecutive years. The magnitude of drought impacts is closely related to the timing of the onset of the precipitation shortage, its intensity, and the duration of the event. The 5-year (1979-83) drought in northeast Brazil is a good case in point. In this series of years, 1979 and 1980 were both drought years in the classic sense (i.e., a significant deficiency during the principal rainy reason). In 1981, the seasonal rainfall totals were slightly above normal but the temporal distribution resulted in agricultural drought. In 1982, the opposite pattern occurred (meteorological drought), and the results were less adverse for agriculture. These 4 "drought" years were followed by the most severe drought year (1983) of the previous 25 years (Magalhães et al., 1988).

Droughts also differ in terms of their spatial characteristics. The areas affected by severe drought evolve gradually, and regions of maximum intensity shift from season to season. In larger countries, such as Brazil, China, India, the United States, or Australia, drought would rarely, if ever, affect the entire country. During the severe drought of the 1930s in the United States, for example, the area affected by severe drought never exceeded 65% of the country (see Figure 1). In India, the droughts of this century have rarely affected more than 50% of the country. An exception occurred in 1918-19, when 73% of the country was affected (Sinha et al., 1987). On the other hand,

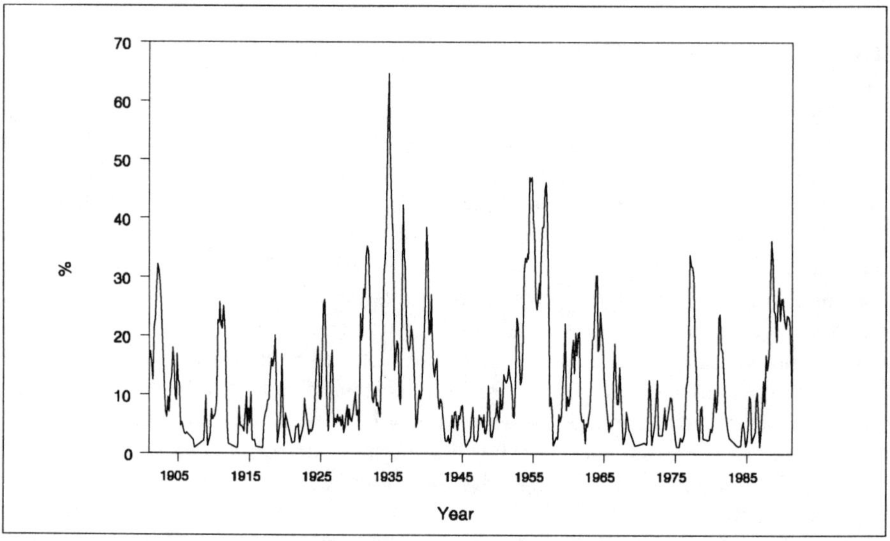

Figure 1. Percent area of the United States experiencing drought, 1895-1991. Compiled from data provided by the National Climatic Data Center, National Oceanic and Atmospheric Administration, Asheville, North Carolina, U.S.A.

it is indeed rare for drought *not* to exist in a portion of these countries in every year. For example, Figure 1 illustrates that in the United States the percent area affected by drought is often greater than 10%. Thus, the governments of these larger countries are more accustomed to dealing with water shortages and have established an infrastructure to respond, albeit reactively. For smaller countries, it is more likely that the entire country may be affected since droughts are usually regional phenomena—they result from large-scale anomalies in atmospheric circulation patterns that become established and persist for periods of months, seasons, or longer.

From a planning perspective, the spatial characteristics of drought have serious implications. Nations should know the probability that drought may simultaneously affect all or several major crop-producing regions within their borders and develop contingencies if such an event were to occur. Likewise, it is important to know the chances of a regional drought simultaneously affecting agricultural productivity in their country as well as adjacent or nearby nations on whom they are dependent for food supplies. In some instances, a nation's primary drought mitigation strategy may be to import food from nearby nations, ignoring the likelihood that a drought may have significant regional impacts. Likewise, the occurrence of drought worldwide or in the principal grain exporting nations, such as occurred during the ENSO event of 1982-83 (Glantz et al., 1987; Glantz et al., 1991), may significantly alter a developing country's access to food from donor governments.

IMPACTS OF DROUGHT

The impacts of drought are diverse and often ripple through the economy. Thus, impacts are often referred to as direct or indirect, or they are assigned an order of propagation (i.e., first-, second-, or third-order) (Kates, 1985). Conceptually speaking, the more removed the impact from the cause, the more complex its link to the cause. In other words, a loss of yield resulting from drought is a direct or first-order impact of drought. However, the consequences of that impact (e.g., loss of income, farm foreclosures, outmigration, government relief programs) are secondary or tertiary impacts. First-order impacts are usually of a biophysical nature while higher-order impacts are usually associated with socioeconomic valuation, adjustment responses, and long-term "change." Asfaw (1989) succinctly summarized drought impacts as "direct or indirect, either singular or cumulative, immediate or delayed."

Because of the number of affected groups and sectors associated with drought, the geographic size of the area affected, and the difficulties connected with quantifying environmental damages and personal hardships, the precise determination of the financial costs of drought is an arduous task. Average annual estimates of the direct losses attributable to drought are misleading. Although some drought-related costs and losses may occur each year in some countries, in most instances they tend to occur in clusters around major single- or multiple-year events. Therefore, direct and indirect losses may be extremely large for one or two consecutive years and then negligible for several years. This clustering of drought-related costs and losses is repeated over and over again. For example, northeast Brazil experienced a severe drought from 1979 to 1983 that was preceded and followed by a series of favorable or wet years. The same can be said for Kenya, 1983-84 (Downing et al., 1987); Zimbabwe, 1981-84 (Makarau and Marume, 1989); Botswana, 1979-80 (Moremi, 1987); and India, 1980-82 (Sinha et al., 1987) and 1988-89 (Venkateswarlu, 1992). The ebb and flow of dry and wet years (and thus the drought-related costs and losses) hinders the preparedness process in all countries. Human nature is to assume that next year will be a "good" year.

The impacts of drought can be classified into three principal sectors: economic, environmental, and social. Table 1 illustrates the principal impacts associated with each of these sectors. The economic impacts of drought are numerous, ranging from direct losses in the broad agricultural and agriculturally related sectors, including forestry and fishing, to losses in recreation, transportation, banking, and energy. Other economic impacts would include added unemployment, increases in food prices and overall disruption of food supply, strain on financial institutions because of farm foreclosures, increased costs of new or supplemental water resource development, and loss of revenue to local, state, and federal government. Environmental losses are the result of damages to plant and animal species, wildlife habitat, and air and water quality; forest and range fires; degradation of landscape quality; and soil erosion. These losses are difficult to quantify, but growing public awareness and concern for environmental quality has forced public officials to focus greater attention on this problem. Increasing levels of environmental regulation (e.g., water quality, preservation of wildlife habitat) have imposed a new layer of constraints on water managers during water-short periods. This trend is likely to continue. Social impacts mainly involve public safety, health, conflicts between water users, inequities in the distribution of impacts and disaster relief programs, loss of life, increased social unrest, depopulation of rural areas, and reduced quality of life.

SUMMARY

Drought is, indeed, a complex and poorly understood phenomenon that affects more people than any other natural hazard. Impacts are far-reaching and may linger for months or even years beyond the termination of the event. The economic, social, and environmental impacts of drought result from complex interactions between physical and social systems that are difficult to quantify. Scientists and policy makers must understand the characteristics of drought and appreciate the magnitude and complexity of impacts in order for viable assessment and response strategies to be established. The aim of these strategies is to reduce societal vulnerability to periods of water shortages.

Problem Sectors	Impacts
Economic	· loss from crop production annual and perennial crop losses; damage to crop quality reduced productivity of cropland (wind erosion, etc.) insect infestation plant disease wildlife damage to crops · loss from dairy and livestock production reduced productivity of rangeland forced reduction of foundation stock closure/limitation of public lands to grazing high cost/unavailability of water for livestock high cost/unavailability of feed for livestock high livestock mortality rates increased predation range fires · loss from timber production forest fires tree disease insect infestation impaired productivity of forest land · loss from fishery production damage to fish habitat loss of young fish due to decreased flows · loss of national economic growth, retardation of economic development · income loss for farmers and others directly affected · loss from recreational businesses · loss to manufacturers and sellers of recreational equipment · increased energy demand and reduced supply because of drought-related power curtailments · costs to energy industry and consumers associated with substituting more expensive fuels (oil) for hydroelectric power · loss to industries directly dependent on agricultural production (e.g., machinery and fertilizer manufacturers, food processors, etc.) · decline in food production/disrupted food supply increase in food prices increased importation of food (higher costs) · unemployment from drought-related production declines · strain on financial institutions (foreclosures, greater credit risks, capital shortfalls, etc.) · revenue losses to federal, state, and local governments (from reduced tax base) · revenues to water supply firms revenue shortfalls windfall profits

Table 1. Classification of drought-related impacts (modified from Wilhite, 1993).

Problem Sectors	Impacts
	· loss from impaired navigability of streams, rivers, and canals · cost of water transport or transfer · cost of new or supplemental water resource development
Environmental	· damage to animal species wildlife habitat lack of feed and drinking water disease increased vulnerability to predation (e.g., from species concentration near water) · wind and water erosion of soils · damage to fish species · damage to plant species · water quality effects (e.g., salt concentration) · air quality effects (dust, pollutants) · visual and landscape quality (dust, vegetative cover, etc.)
Social	· food shortages (decreased nutritional level, malnutrition, famine) · loss of human life (e.g., food shortages, heat) · public safety from forest and range fires · conflicts between water users · health-related low flow problems (e.g., diminished sewage flows, increased pollutant concentrations, etc.) · inequity in the distribution of drought impacts/relief · decreased living conditions in rural areas · increased poverty · reduced quality of life · social unrest, civil strife · population migration (rural to urban areas)

Table 1 (contd.).

ABOUT THE AUTHOR

Donald A. Wilhite is professor of agricultural climatology in the Department of Agricultural Meteorology and director of the International Drought Information Center at the University of Nebraska-Lincoln. He specializes in studies of the impact of climate on society and societal response to climatic events, particularly drought. Dr. Wilhite is coeditor of *Planning for Drought: Toward a Reduction of Societal Vulnerability*, published by Westview Press in 1987. He has recently written a guidebook on drought preparedness for developing countries under sponsorship of the United Nations Environment Program. Dr. Wilhite is chair of the Committee on Applied Climatology of the American Meteorological Society.

REFERENCES

American Heritage Dictionary. 1976. Houghton Mifflin, Boston.

Asfaw, D. 1989. Drought preparedness in Ethiopia. Unpublished paper. Drought Management and Preparedness Training Seminar for Eastern and Southern Africa. Gaborone, Botswana.

Coughlan, M. J. 1987. Monitoring drought in Australia. *In* D. A. Wilhite and W. E. Easterling, eds. *Planning for Drought: Toward a Reduction of Societal Vulnerability*; Chapter 10. Westview Press, Boulder, Colorado, U.S.A.

Downing, T. E.; K. W. Gitu; and C. M. Kamau, eds. 1987. *Coping with Drought in Kenya: National and Local Strategies*. Lynne Rienner Publishers, Boulder, Colorado, U.S.A.

Dracup, J. A.; K. S. Lee; and E. G. Paulson, Jr. 1980. On the definition of droughts. *Water Resources Research* 16(2):297-302.

Gibbs, W. J.; and J. V. Maher. 1967. *Rainfall Deciles as Drought Indicators*. Bureau of Meteorology Bulletin No. 48. Melbourne, Australia.

Glantz, M. H.; and R. W. Katz. 1977. When is a drought a drought? *Nature* 267:192-93.

Glantz, M. H.; R. W. Katz; and M. Krenz. 1987. *Climate Crisis: The Societal Impacts Associated with the 1982-83 Worldwide Climate Anomalies*. National Center for Atmospheric Research, Boulder, Colorado, U.S.A.; and U.N. Environment Program.

Glantz, M. H.; R. W. Katz; and N. Nicholls. 1991. *Teleconnections Linking Worldwide Climate Anomalies*. Cambridge University Press, Cambridge.

Hagman, G. 1984. *Prevention Better than Cure: Report on Human and Natural Disasters in the Third World*. Swedish Red Cross, Stockholm.

Jose, A. M.; F. O. Magnayon; and F. D. Hilario. 1991. Climate impact assessment for agriculture in the Philippines. Unpublished paper. National Workshop on Drought Planning and Management in the Philippines. Quezon City, Philippines.

Kates, R. W. 1985. The interaction of climate and society. *In* R. W. Kates, J. H. Ausubel, and M. Berberian, eds. *Climate Impact Assessment*; Chapter 1. John Wiley and Sons, New York.

Klemeš, V. 1987. Drought prediction: A hydrological perspective. In D. A. Wilhite and W. E. Easterling, eds. *Planning for Drought: Toward a Reduction of Societal Vulnerability*; Chapter 7. Westview Press, Boulder, Colorado, U.S.A.

Lee, D. M. 1979. Australian drought watch system. *In* M. T. Hinchey, ed. *Botswana Drought Symposium*; pp. 173-87. Botswana Society, Gaborone, Botswana.

Magalhães, A. R.; H. C. Filho; F. L. Garagorry; J. G. Gasques; L. C. B. Molion; M. da S. A. Neto; C. A. Nobre; E. R. Porto; and O. E. Rebouças. 1988. The effects of climatic variations on agriculture in Northeast Brazil. *In* M. L. Parry, T. R. Carter, and N.

T. Konijn, eds. *The Impact of Climatic Variations on Agriculture*. Vol. 2, *Assessments in Semi-Arid Regions*; pp. 273-380. Kluwer Academic Publishers, Boston.

Makarau, A.; and W. Marume. 1989. Drought management in Zimbabwe: The 1981-85 experience. Unpublished paper. Drought Management and Preparedness Training Seminar for Eastern and Southern Africa. Gaborone, Botswana.

Moremi, T. C. 1987. Drought planning and response: Botswana experience. *In* D. A. Wilhite and W. E. Easterling, eds. *Planning for Drought: Toward a Reduction of Societal Vulnerability*; Chapter 26. Westview Press, Boulder, Colorado, U.S.A.

Olidapo, E. O. 1985. A comparative performance analysis of three meteorological drought indices. *Journal of Climatology* 5:655-64.

Palmer, W. C. 1965. *Meteorological Drought*. Research Paper No. 45. U.S. Weather Bureau. Washington, D.C.

Palmer, W. C. 1968. Keeping track of crop moisture conditions, nationwide: The new crop moisture index. *Weatherwise* 21(4):156-61.

Sandford, S. 1979. Towards a definition of drought. *In* M. T. Hinchey, ed. *Botswana Drought Symposium*. Botswana Society, Gaborone, Botswana.

Sinha, S. K.; K. Kailasanathan; and A. K. Vasistha. 1987. Drought management in India: Steps toward eliminating famines. *In* D. A. Wilhite and W. E. Easterling, eds. *Planning for Drought: Toward a Reduction of Societal Vulnerability*; Chapter 27. Westview Press, Boulder, Colorado, U.S.A.

Subrahmanyam, V. P. 1967. *Incidence and Spread of Continental Drought*. WMO/IHD Report No. 2. WMO, Geneva.

Tannehill, I. R. 1947. *Drought: Its Causes and Effects*. Princeton University Press, Princeton, New Jersey, U.S.A.

Venkateswarlu, J. 1992. Disaster management: A national perspective. *Drought Network News* 4(1):4-6. International Drought Information Center, University of Nebraska, Lincoln, Nebraska, U.S.A.

Whipple, W., Jr. 1966. Regional drought frequency analysis. *Proceedings of the American Society of Civil Engineers* 92(IR2):11-31.

Wilhite, D. A.; and M. H. Glantz. 1985. Understanding the drought phenomenon: The role of definitions. *Water International* 10:111-20.

Wilhite, D. A.; N. J. Rosenberg; and M. H. Glantz. 1986. Improving federal response to drought. *Journal of Climate and Applied Meteorology* 25:332-42.

Wilhite, D. A. 1992. Drought. *Encyclopedia of Earth System Science*, Vol. 2; pp. 81-92. Academic Press, San Diego, California, U.S.A.

Wilhite, D. A. 1993. *Preparing for Drought: A Guidebook for Developing Countries*. United Nations Environment Program, Nairobi, Kenya (in press).

World Meteorological Organization. 1975. *Drought and Agriculture*. WMO Technical Note No. 138. Report of the CAgM Working Group on the Assessment of Drought. WMO, Geneva.

Yevjevich, V. 1967. *An Objective Approach to Definitions and Investigations of Continental Hydrologic Droughts*. Hydrology Papers No. 23, Colorado State University, Fort Collins, Colorado, U.S.A.

Chapter 2
Monitoring Regional Drought Conditions

Kenneth G. Hubbard

INTRODUCTION

Drought is a recurring phenomenon that receives greatest regional attention when it makes a major impact on the economy. Here an impact is taken to mean any change that can be directly or indirectly attributed to a drought. Drought intensity is closely tied to changes in the hydrological cycle because drought impacts can occur at any juncture in the cycle where natural or managed systems are sensitive to changes in precipitation, snowpack, streamflow, surface water storage, soil water storage, or groundwater.

One impact important to agriculture is the unusual reduction of soil water, a critical component of the hydrologic cycle for crops, during a drought. Reduced soil water can cause reduced crop productivity, especially when accompanied by dry, hot atmospheric conditions. An example of an indirect drought impact is the reduction of barge traffic associated with low river flows induced by drought. Drought impact is usually referred to as a *stressor* on natural or managed environments. Some impacts can be positive, at least in the short term. For example, drought conditions at the time of crop planting or harvesting may be ideal for field operations. Days available for the construction industry also are generally higher during a drought. For the most part, we take drought impacts to be negative, as is the case when drought causes crop stress, restrictions on irrigation, or the need for more frequent irrigations.

It has been tacitly assumed in the study of drought and drought impacts that better data will result in more accurate information and that the best information is needed to make wise decisions during a drought. This assumption has guided the preparation of this chapter.

The purpose of this chapter is to discuss drought monitoring from a physical standpoint, assuming that monitoring is an attempt to follow water through natural and managed systems. The intent is not to promote physical monitoring above the need to identify secondary impacts, although any prediction of the latter is certainly dependant on a realization of the former. In the past decade, advances in monitoring technology have improved our capability to provide early warning of drought in a region. A comprehensive monitoring system will encompass both field measurement and computer modeling components. The exact form of the components making up a monitoring system will be determined by the primary drought impacts. The drought impacts are governed by the social, economic, and environmental characteristics of a region.

The following section discusses the importance of certain variables for drought monitoring, gives a brief indication of how each is measured, and provides examples, where appropriate, of agencies responsible for monitoring these variables in the United States. To monitor all components of water in the hydrologic cycle would require tracking of the water in the atmosphere overhead; in the glaciers, icepacks, snowpacks, streams, rivers, and open waters of lakes and ponds; in the soils that support surface vegetation; in the groundwater; and in the oceans.

The section discussing a regional drought monitoring system is included as an example of a specific system developed around the soil water variable. The final section, Forecasts and Outlooks, is included because it is essential to know when a drought may lift its hold on a region and thereby allow resumption of normal activities.

MONITORING DROUGHT: THE WATER-RELATED VARIABLES

In the United States and elsewhere, monitoring systems have been developed by federal and provincial government agencies to track various components of the hydrologic or water cycle. In many instances, the data/informational products from these systems may be readily available in reports and releases. Alternatively, one might find such publications in a library. In the United States, many agencies are beginning to offer dial-up computer systems so that interested persons have immediate access to the most recent information. Typically, a computer information system offers the same information that will appear in subsequent publications of the agency, but the information is available before publications are printed, albeit in preliminary form with regard to accuracy and quality control. The examples used throughout this section are specific to the collection and dissemination of drought-related information for the United States. Similar agencies are at work in other countries, but because of the space limitations of this chapter, their contributions will not be included. It is best to check with local officials in the region of interest for information of the type discussed herein.

In the discussion that follows, some of the key variables in drought monitoring are discussed, along with the data collection techniques used in the United States. Of course, the variables of importance vary from region to region. Therefore, in establishing a monitoring or early warning system for drought, consideration must be given to the most appropriate variables for that climatic regime while taking into account the economic, social, and environmental impacts that characterize that region.

Atmosphere

It is important to monitor the water present in the atmosphere because it is the source of moisture involved in cloud formation and precipitation processes. This moisture, which occurs in the form of water vapor, droplets, ice particles, and so forth, is a key variable in the computer models that forecast precipitation. Several data collection systems are used.

One system relies on a number of balloon soundings (radiosondes) to define the structure of important variables in the atmosphere above major cities. The temperature, pressure, humidity, windspeed, and direction are determined from the soundings. This information is used to initialize weather forecast models so that the future state of the atmosphere can be predicted.

The next generation of weather radars for operational use are now being installed by the National Weather Service (NWS) and will provide digital data for the identification of cloud water content and associated falling raindrops or ice particles. Identification of water in this form will be possible on a dense grid (4 km resolution) for nearly the entire United States. The NWS is responsible for these networks, and the National Climatic Data Center (NCDC) is responsible for archiving the data.

Snowpack

Low snowpack in mountain regions is an early indicator of dry local conditions and possible drought impacts at lower elevations. Low snowpack usually means contributions to streamflow from snowmelt will be reduced, thus reducing streamflow. This in time can affect drinking water supplies for communities at lower elevations that are dependent on streamflow and/or surface water storage. Drought conditions in the mountains may lead to high fire danger. Snowpack is measured on a number of established snow courses by measuring the depth of snow cores and the water content as a function of depth into the snowpack. Snowpack measurements can also be made by using snow pillows, winterized rain gauges, or ultrasonic snow depth sensors. Output from any of these sensors can be collected through satellite, meteor burst, or ground-based radio frequency communications. Snowpack information is compiled by the Soil Conservation Service (SCS), U.S. Department of Agriculture. The SCS publishes the *Water Supply Outlook Report* for the western United States and the *Basin Outlook Reports*. These reports indicate the current season's snowpack in comparison to minimum, average, and maximum snowpack for the period of record. SCS also offers computer access to a system called the Centralized Forecasting System (CFS). The CFS provides information that is typically updated five working days after the first of the month. The on-line information is provided in the same format as the published reports.

Streamflow

Streamflow measurements are important for a number of reasons, all having to do with the use of water directly from streams and rivers by humans or other life forms. Water diverted from streams and rivers for power generation, irrigation, or industrial or city use is important to the well-being of a community. Streamflow measurements determine the volume of water passing a given point along a stream or river in a unit of time. Various federal and state agencies maintain streamflow gauges, and they make use of the data in a number of ways. The SCS includes streamflow in their reports. The U.S. Army Corps of Engineers (COE) makes use of streamflow information to determine the runoff; this data helps COE effectively manage discharge from reservoirs. The COE makes data available in monthly news releases, and state water agencies also keep records of streamflow and publish updates. The NWS River Forecast Centers maintain data on streamflow and use basin simulation models to provide flood warnings, but flood danger is usually lessened during drought periods.

Precipitation

Precipitation is of course that portion of the atmospheric water that actually falls on the earth's surface. Its fate is to either become runoff or enter the soil. Runoff feeds the streams and lakes, while soil water can either eventually become part of the groundwater or be taken up by plant roots and used in evaporation. Precipitation is important for all the reasons that streamflow, soil water, and groundwater are important. Precipitation is usually measured as the amount of water caught in a cylindrical rain gauge. Precipitation gauges are a major component of networks operated by the NWS, Federal Aviation Agency, Bureau of Land Management, National Forest Service, state agencies, and local agencies. Monthly reports for selected stations are published by the National Climatic Data Center for each state. Subscriptions provide high-quality data, but published reports are not available until two to three months after the fact. More timely NWS reports, by computer, can be obtained by subscribing to one of the many vendors in the private sector now providing NWS data on a real-time basis. Other sources for reports that are based on preliminary data are state climatologists (SCs), Regional Climate Centers (RCCs), and the Climate Analysis Center (CAC). During a drought, many of these climate offices prepare special drought advisories specific to their area of responsibility. In addition, many of them offer computer dial-up systems on which preliminary data is available.

The NEXRAD radar system in the United States offers potential as a precipitation measuring system. Although challenges remain with regard to relating the radar reflectivity factor to the rainfall rate in real time, this appears to be the first major opportunity for advances in spatial resolution of precipitation monitoring since completion of the first order and cooperative weather station networks.

Reservoir Storage

Water shortages in reservoirs during drought can reduce water available for irrigation, power generation, recreation, municipal uses, industrial uses, river transportation, and minimum in-stream flows. Therefore, depending on the use of individual reservoirs, it is usually very important to monitor the water held in storage. Available water is determined by measuring the reservoir water level and employing its relationship to volume of water held by the reservoir. Reservoir levels are monitored by the agencies responsible for management of the reservoir storage and discharges. For reservoirs of moderate to large size, monthly reports are generally available from the responsible agency (such as the COE and the Bureau of Reclamation).

Groundwater

Groundwater is the source of drinking water for many cities, and some industries rely on groundwater as well. It is thus very important to track the groundwater. Water quality can decrease as the groundwater level decreases, so it is important to monitor shallow systems more frequently. Deeper groundwater systems respond more slowly to drought because of the large distance separating dry conditions at the surface and the water table. For this reason, water levels may only be read a few times during the course of a year. Water levels in groundwater wells are tracked by municipalities and state and federal government agencies. Groundwater reports are available from the United States Geological Survey and from various state water agencies.

Soil Water Reserves

Adequate soil water is necessary for achieving high levels of crop production. Soil water can be measured by a variety of techniques (Schmugge et al., 1980; Hanks, 1987). Soil water status is rated in qualitative terms in the weekly reports of the state agricultural statistics offices. This rating does not reflect regular measurements from fixed stations in a continuing network, nor do such networks exist except on an exploratory basis. Both the Midwest Climate Center (MCC) and the High Plains Climate Center (HPCC) use soil water balance models to estimate soil water reserves within their region. The case study given later in this chapter is a discussion of the HPCC soil water monitoring system.

Evaporation and Evapotranspiration

Evaporation from soil and water surfaces and transpiration from plant surfaces are the two processes responsible for transferring water from the surface back into the atmosphere. It is important to know the magnitude of these terms in order to know the degree of demand placed on soil water reserves. In the case of plants, the demand can also be an indication of plant water stress during drought. Several techniques are available for measuring these variables (Rosenberg et al., 1983). In the section on regional drought monitoring, an approach to estimation of evaporation and transpiration in cropped areas is discussed.

Drought Indices

The term *drought index* refers to a derived variable, the value of which reflects the severity of a drought. The Palmer Drought Severity Index (PDSI) is calculated from temperature and precipitation values, and it ranges from extreme drought (-4.00) to excessively wet (+4.00), although values of the PDSI occasionally reach -8.00. The Crop Moisture Index (CMI) is a related index that gives the status of the current moisture compared to moisture demand. These indices are available in the *Weekly Weather and Crop Bulletin*, published jointly by the U.S. Department of Agriculture and the NWS. These indices of drought are among the most well known and are widely used. The PDSI and CMI are essentially indicators of climatological drought and were not intended for use in analyzing climate impacts on specific activities. Where a specific type of impact is to be examined, it is suggested that more specific indices be used.

The Surface Water Supply Index (SWSI) has been derived for use in the mountainous regions where the PDSI did not prove effective because of its tendency to indicate extreme drought long after dry conditions had ended. The SWSI is a weighted index, derived for each major river basin, that generally expresses the potential availability of the forthcoming season's water supply. Inputs to the SWSI are reservoir storage, snowpack water equivalent, and precipitation. The SWSI ranges from -4.00 to +4.00 to reflect extremely poor to plentiful water supply prospects, respectively. SWSI is available in the *Colorado Water Supply Outlook Report*.

A new drought index has been developed for corn (Meyer et al., 1992a and 1992b). This index is derived from inputs of temperature, humidity, windspeed, solar radiation, and precipitation. The crop specific drought index (CSDI) values range from zero to one as the predicted production ranges from crop failure to bumper crop. The High Plains Climate Center continues to develop CSDI for other crops. Outlooks for both CSDIC (corn) and CSDIS (sorghum) can be prepared at any time in the growing season by using various weather scenarios for the remainder of the growing season. Using historical records, an "end of season" CSDI estimate can be assigned an appropriate probability level.

The National Oceanic and Atmospheric Administration (NOAA) satellites also offer some unique surface monitoring capabilities. NOAA Advanced Very High Resolution Radiometer (AVHRR) data has been used to form the normalized difference vegetation index (NDVI). The NDVI is simply the difference in radiation measured in two different wavelength bands (0.55 to 0.68 m, and 0.73 to 1.1 m) divided by the sum of the radiation in these two bands. The monthly NDVI can be mapped and used as an estimate of the intercepted photosynthetically active radiation. It has been suggested that the maximum NDVI for a given season can serve as an indicator of drought severity by inferring deficiencies in photosynthetic capacity in drought years as contrasted with other years (Tucker and Goward, 1987). Chapter 3 provides a more detailed discussion of the application of this technology to drought monitoring.

A REGIONAL DROUGHT MONITORING SYSTEM

As we prepare to deal with drought issues in the "information age," it is good to review some points related to drought monitoring. If new monitoring must be undertaken, it is important to develop a monitoring plan. This should include a justification or rationale for measurements to be taken and an error analysis, if possible, to decide what accuracy is needed to accomplish the goals of monitoring.

Technological advances in communication, computers, and remote sensing have greatly increased our ability to monitor the important characteristics and impacts of weather. The use of electronic equipment to automate the collection of measurements from weather-related sensors at remote sites has increased our ability to collect weather data (Hubbard et al. 1983; Hubbard, 1987). Such advances in data collection are finding their way into the NWS program of modernization, as more than 1,000 new stations are planned for the Automated Surface Observing System (ASOS, 1988). Below is a discussion of the measurement and modeling components of a regional drought monitoring system.

Automated Weather Data Network (AWDN)—A Regional Monitoring Tool

The Automated Weather Data Network (AWDN) in Nebraska was initiated in 1981. Information from the network is used in operational decisions that depend on dynamic and ever-changing factors or variables that describe the most recent weather (Hubbard et al., 1983) in a region where irrigation of agricultural crops is widely practiced. The network demonstrated that accurate and timely weather data could be used to calculate water use and thus determine, in a scientifically supported manner, the

Sensor	Variable	Installation Ht.	Accuracy	Hourly
Thermistor	Air temp.	1.5 m	0.25 C	Avg.(C)
Thermistor	Soil temp.	-10.0 cm	0.25 C	Avg.(C)
Si Cell Pyranometer	Radiation —Global	2.0 m	2%	Flux (W m^{-2})
Cup Anemometer	Wind speed	3.0 m	5% 0.5m/s start-up	Total passage (ms^{-1})
Wind Vane	Wind direction	3.0 m	2°	Vector direction
Coated Circuit	Rel. humidity	1.5 m	5%	Avg. (%)
Tipping Bucket	Precipitation	0.5-1.0 m	5%	Total (mm)

Table 1. Variables measured by the AWDN.

timing of irrigations. This type of irrigation scheduling minimizes energy/water use while maintaining high production levels.

The AWDN functions to record hourly information from sensors at each station. The variables measured are shown in Table 1. The installation heights shown are standard for AWDN stations (Meyer and Hubbard, 1992).

Much of the early growth of this regional network was due to the interest of researchers who were operating digital weather stations without the benefits of telecommunication or a data management system. As time passed, private sector

Year	No. of Stations	Year	No. of Stations
1981	5	1987	49**
1982	14	1988	60
1983	21	1989	74
1984	29	1990	83
1985	47	1991	91
1986	51	1992	93

**The total number of stations declined in 1987 because a number of stations were removed from service.

Table 2. Number of stations in AWDN, by year.

interests expressed a desire to be a part of the network. Resource management agencies also have taken an active role in adding and supporting stations in the network. One unique class of weather station sponsor is the community consortium. In this case, a number of interested parties from a community (e.g., agrichemical dealers, farm elevators, radio station, public power agency) decide to share in the expense of purchasing and maintaining a station. The yearly growth of the AWDN is shown in Table 2.

In 1983, the AWDN began to grow into surrounding states. Currently, the 93 stations in the regional network are distributed as follows: Colorado, 4; Iowa, 8; Kansas, 16; Missouri, 2; Nebraska, 38; North Dakota, 14; South Dakota, 9; and Wyoming, 2. In general, each state is responsible for maintaining its weather stations. The states with larger numbers of stations run a near-real time network to serve clientele within their boundaries.

In the High Plains network, about 1 Mb of data is produced annually for any three stations. However, if this data is to be used effectively, it must be easy to access. Thus, data management is a real concern. The approach has been to develop a data management system written entirely in FORTRAN (Hubbard et al., 1992).

The cost of a station is about $4,000. Average annual costs associated with the network include: local telephone service ($480), telephone calls ($180), travel ($200), repair costs ($100), replacement costs ($100), and labor ($1,250). The total cost here is $2,310 per year, but costs may vary with the number of stations that are operating and other local rates.

A Regional Soil Water Model

Soil water content is the most variable of all the resources for agricultural cropping systems. Until recently, models had not been widely used to estimate the water in the root zone on a wide-area basis.

Applications of models to soil water estimation under a variety of crops have recently been reported in the literature (e.g., Kunkel, 1990; Robinson and Hubbard, 1990; Steiner et al., 1991; and Arkin and Dugas, 1982). These models make use of climatological methods to estimate potential evapotranspiration (ET_p). A discussion of such methods for estimating ET_p is presented by Rosenberg et al. (1983). The methods require one or more of the ambient weather measurements of such variables as solar radiation, windspeed, temperature, and humidity. The resulting estimates of potential evapotranspiration are only a function of existing weather and can be considered as an atmospheric demand placed on growing plants.

Monteith (1963) presented a method to estimate evaporation from surfaces at optimal (ET_p) or limited water supply (ET) based on the vapor pressure gradient (leaf minus air) and two resistance terms—one describing the resistance to vapor escaping from the leaf and the other an aerodynamic resistance. Steiner et al. (1991) used leaf area index to estimate canopy resistance from well-watered sorghum and found that the resistance method of calculating potential evapotranspiration gave closest agreement to lysimeter measurements. The following approach (Hubbard, 1992) is used to calculate evapotranspiration in the High Plains Soil Water Model:

$$\rho_w L_v ET = [\Delta(R_n+G) + \rho C_p (e_s-e_a)/r_a] / \{\Delta + \gamma(r_a + r_c)/r_a\}$$

The terms are net radiation (R_n), soil heat flux (G), saturated vapor pressure (e_s), and vapor pressure of air (e_a). Other terms in the equation are the psychrometric constant (γ) and slope of the saturation vapor pressure curve (Δ) with respect to temperature. The symbols ρ_w and L_v are the density of liquid water and the latent heat of vaporization, respectively. The terms ρ and C_p refer to the density and specific heat capacity, respectively, of air. For R_n and G in units of $MJ/m^2/day$, ρ_w in kg/m^3, and L_v in MJ/kg, the units of ET are m/day. The variables r_a and r_c are the aerodynamic and canopy resistances. Further details are given in the appendix at the end of this chapter.

The soil water (S) then changes with time as follows:

$$dS/dt = P - ET - D_r - R_o$$

where P is the precipitation, ET the evapotranspiration, D_r the drainage below the root zone, and R_o the runoff at the surface.

Performance of the model described by Robinson and Hubbard (1990) was reported for 20 separate examples involving five crops, nine locations, and two years of data in the High Plains region of the United States. The root mean square error (RMSE) for the total soil water in the root zone was found to range from 13 to 86 mm, with the majority of RMSE ranging from 14 to 46 mm. For the most part, the variance explained by the model was in excess of 70%. The sites involved in this study were all in Nebraska, South Dakota, or Wyoming and varied widely with respect to timing and amount of precipitation, range of temperature, and humidity values.

Local and Regional Use of the Soil Water Model

The model is used to simulate soil water at sites where measurements of the same are available. This provides a useful validation of the model, and good agreement increases confidence that the model is performing well.

The weather variables are available at many other sites in the AWDN. Soils information is also available in various soils maps. The agricultural soil representing the greatest area in the vicinity of these weather stations is used to specify field capacity and wilting point as well as percent clay, silt, and loam.

By checking planting progress in the *Weather and Crop Reports* and with local extension personnel, an average planting date can also be estimated. With this information, soil water for various crops is simulated at each site in the AWDN.

During the growing season, this simulation usually is current through the most recent day. Individual values are plotted according to the latitude and longitude of the location, and contour maps are drawn by the Kriging technique. One example of the type of map that can be produced during the growing season is shown here for corn over a portion of the High Plains region.

FORECASTS AND OUTLOOKS

Forecasts can be important to decision makers, especially during a drought. Forecasts can indicate the most likely temperature and precipitation conditions for the immediate future and thus the likelihood of relief from existing drought conditions. The NWS provides several forecasts to cover different time periods. The daily forecast covers the remainder of the current day and all of the next day; it is updated several times each day. The extended forecast covers the period 3-5 days in the future. The 6-10 day forecast is the longest period into the future covered by conventional forecasting techniques. Long-range forecasting techniques are used by CAC to produce 30- and 90-day outlooks on the 1st and 15th of each month.

As we mentioned previously, outlooks for drought index values can be given on a probabilistic basis. In many of the examples given, this is the case. In the event of an El Niño/Southern Oscillation (ENSO), CAC prepares an advisory to discuss the ENSO and any possible effects of ENSO on future weather.

CONCLUSIONS AND RECOMMENDATIONS

The preceding examination of the water-related data/information collected shows that the data characteristically resides with many agencies of provincial and federal government. One conclusion that can be drawn is that the dispersed and complex nature of data collection systems should lead those investigating the status of drought to involve climate expertise.

Further, sweeping developments in technology are changing the various networks that are used to monitor surface and atmospheric conditions. Users should anticipate the development of drought monitoring indexes that are more specific to the impacts in a region.

APPENDIX—SOIL WATER MODEL DEFINITIONS

The following equations for aerodynamic and canopy resistances were derived.

$$r_a = \rho C_p / \{\gamma f(U)\}$$

$$r_c = r_a \{(\Delta + \gamma)/\gamma\} (1 - f_v f_u f_s K_c) / \{f_v f_u f_s K_c\}$$

where ρ and C_p are the density and specific heat capacity of dry air. The wind function is the same as that from Robinson and Hubbard (1990): $f(U) = 7.1 + 0.068\ U$. When U is expressed in km/day, the wind function has units of MJ (KPa m^2 day)$^{-1}$. Here K_c is a crop coefficient, determined from independent experiments by examining the ratio of T to ET_p in various growth stages for a well-watered crop. Growth stage is estimated according to the accumulation of growing degree days (base 10° and upper limit 30°C for a warm season crop). K_c values are also specified according to the accumulation of growing degree days and range from 0.0 at emergence to slightly more than 1.0 at the end of the vegetative stage.

The factor f_s allows the model to simulate transpiration when the crop is not well watered and is referred to as a soil water reduction factor. In the model, f depends on current soil water status:

$$f_s = 1.0 \quad \text{if } S/AW_p > F \text{ and}$$
$$f_s = S/(F \cdot AW_p) \quad \text{if } S/AW_p \leq F$$

The factor f_s is equal to one (1.0) at high soil water (S) contents relative to the potential available water in the soil (AW_p). F is a threshold value and was found to range from 0.35 to 0.6 at different sites (Robinson and Hubbard, 1990).

The two atmospheric reduction factors are calculated in the following manner:

$$f_U = 1.0 \quad \text{if } U < U_1 \text{ and}$$
$$f_U = 1. - (U-U_1)/(U_2-U_1) \quad \text{if } U_1 \leq U \leq U_2$$
$$f_U = 0.01 \quad \text{if } U > U_2$$

The vapor pressure deficit reduction factor was calculated as follows:

$$f_v = 1.0 \quad \text{if } d < d_1 \text{ and}$$
$$f_v = 1. - (d-d_1)/(d_2-d_1) \quad \text{if } d_1 \leq d \leq d_2$$
$$f_v = 0.01 \quad \text{if } d > d_2$$

where the term d refers to the vapor pressure deficit and the subscripts 1 and 2 represent threshold values of U and d at which these functional relationships are going to change.

The actual evapotranspiration (ET) is the total of transpiration and evaporation:

$$ET = T + E$$

Evaporation on a given day (E) for the current version of the model is calculated from:

$$E=E_p\,(d_o/d)^{0.5}$$

where the potential soil evaporation (E_p) for that day was taken as ET_p. The variable d_o was set to 1.0 on the day of the most recent wetting, and d is the number of days since the last wetting. The result is that evaporation decreases exponentially from the day of the last wetting.

ABOUT THE AUTHOR

Kenneth G. Hubbard is an associate professor in the Department of Agricultural Meteorology and director of the High Plains Climate Center at the University of Nebraska-Lincoln. He specializes in monitoring and modeling the effects of weather on the biosphere. Dr. Hubbard is coeditor of *Impacts of Climate Change and Variability on the Great Plains* and coauthor of *Automated Weather Station Networks* (in press). He was named to *Who's Who in Technology* and is now serving on the American Society of Agricultural Engineers Committee on Standards for Automated Agricultural Weather Stations. Dr. Hubbard is rapporteur for Weather and the Soya Crop for the World Meteorological Organization.

REFERENCES

ASOS. 1988. *ASOS Progress Report—June*. National Weather Service, Silver Spring, Maryland, U.S.A.

Arkin, G. F.; and W. A. Dugas, Jr. 1982. Evaluation of sorghum production strategies using a crop model. *Agrometeorology of Sorghum and Millet in the Semi-Arid Tropics: Proceedings of the International Symposium*. ICRISAT, Patancheru, India.

Hanks, R. J., ed. 1987. Measurement of soil and plant water status. *Proceedings of an International Conference*, Vol. 1. Department of Soils and Biometeorology, Utah State University, Logan, Utah, U.S.A.

Hubbard, K. G.; J. R. Hines; and D. A. Wood. 1992. *Manual for Installing and Using the AWDN System Software*. High Plains Climate Center Report 92-2. Department of Agricultural Meteorology, University of Nebraska, Lincoln, Nebraska, U.S.A.

Hubbard, K. G.; N. J. Rosenberg; and D. C. Nielsen. 1983. Automated weather data network for agriculture. *Journal of Water Resources Planning and Management* 109:(3)213-22.

Hubbard, K. G. 1987. Surface weather monitoring and the development of drought and other climate information delivery systems. *In* D. A. Wilhite and W. E. Easterling, eds. *Planning for Drought: Toward a Reduction of Societal Vulnerability*; pp. 97-112. Westview Press, Boulder, Colorado, U.S.A.

Hubbard, K. G. 1992. Climatic factors that limit daily evapotranspiration in sorghum. *Climate Research* 2(1):73-80.

Kunkel, K. E. 1990. Operational soil moisture estimation for the Midwestern United States. *Journal of Applied Meteorology* 29:1158-66.

Meyer, S. J.; K. G. Hubbard; and D. A. Wilhite. 1992a. The development of a crop-specific drought index for corn. I. Model development and validation. *Agronomy Journal* (in press).

Meyer, S. J.; K. G. Hubbard; and D. A. Wilhite. 1992b. The development of a crop-specific drought index for corn. II. Applications in drought monitoring and assessment. *Agronomy Journal* (in press).

Meyer, S. J.; and K. G. Hubbard. 1992. Nonfederal automated weather stations and networks in the United States and Canada: A preliminary study. *Bulletin of the American Meteorological Society* 73:449-57.

Monteith, J. L. 1963. Gas exchange in plant communities. *In* L. T. Evans, ed. *Environmental Control of Plant Growth*; pp. 95-112. Academic Press, New York.

Robinson, J. M.; and K. G. Hubbard. 1990. Soil water assessment model for several crops in the High Plains. *Agronomy Journal* 82:1141-48.

Rosenberg, N. J.; B. L. Blad; and S. B. Verma. 1983. *Microclimate: The Biological Environment.* Wiley-Interscience, New York.

Schmugge, T. J.; T. J. Jackson; and H. L. McKim. 1980. Survey of methods for soil moisture determination. *Water Resources Research* 16(6):961-79.

Steiner, J. L.; T. A. Howell; and A. D. Schneider. 1991. Lysimetric evaluation of daily potential evapotranspiration models for grain sorghum. *Agronomy Journal* 83:240-47.

Tucker, C. J.; and S. N. Goward. 1987. Satellite remote sensing of drought conditions. *In* D. A. Wilhite and W. E. Easterling, eds. *Planning for Drought: Toward a Reduction of Societal Vulnerability*; pp. 145-51. Westview Press, Boulder, Colorado, U.S.A.

Chapter 3
The Role of NOAA Satellite Data in Drought Early Warning and Monitoring: Selected Case Studies

Gary E. Johnson, V. Rao Achutuni, S. Thiruvengadachari, and Felix Kogan

INTRODUCTION

Although drought is a phenomenon that dates to prehistoric times, recent technological advances permit evaluation of drought from a new perspective. Satellite remote sensing, the ability to detect the characteristics of features from a distance (i.e., without coming into direct contact with them), is a powerful tool for evaluating the temporal and spatial aspects of drought.

Interest in satellite observation and subsequent evaluation of drought may be attributed to several characteristics of remote sensing. These include the fact that remote sensing provides (1) a unique vantage point, (2) a synoptic view, (3) a permanent record or data archive, (4) extra visual information, and (5) cost effectiveness in many cases. In addition, polar-orbiting satellites provide a regular, repetitive view of nearly all of the surface of the planet.

In recent years, numerous international organizations and national governments have shown an interest in the use of data from the Advanced Very High Resolution Radiometer (AVHRR) sensor aboard the National Oceanic and Atmospheric Administration (NOAA) series of polar-orbiting satellites of the United States. These satellites provide twice-daily coverage of the planet's surface in several spectral channels and spatial resolutions. An archive of these data is available for comparison with conditions dating back to the early 1980s. Further, these data are available at many receiving stations around the world; the data are extremely cost-effective and can be made available in near-real time for the evaluation of current conditions.

Remotely sensed data, especially AVHRR data from the NOAA polar-orbiting satellites, has been used in drought early warning and food security applications in several southeast Asian countries. It is also being applied in an operational context in India and has been evaluated for the increased understanding that the data provides concerning the extent and severity of recent droughts in the United States. In this chapter, scientists with an extensive background in the application of satellite data to the problem of drought assessment report on their recent project work and research. As will be noted, researchers are gaining a better understanding of the drought phenomenon through the use of the modern technology of satellite remote sensing.

METHODOLOGY

Large-area monitoring of vegetation dynamics has been made possible by polar orbiting environmental satellites such as SPOT, LANDSAT, and NOAA. Earth resource satellites such as SPOT and LANDSAT provide high spatial resolution (10-20 m for SPOT; 30-79 m for LANDSAT) data, which find extensive applications in land-resource mapping and inventorying. On the other hand, the NOAA series of satellites provide twice-daily coverage on a near-global basis, albeit at lower spatial resolutions (1-16 km) than either LANDSAT or SPOT. This aspect of twice-daily coverage plus the low cost associated with procuring the data makes the NOAA series of satellites ideal for monitoring global vegetation dynamics.

NOAA has adapted satellite crop-monitoring technology, developed earlier for monitoring large-area crop conditions in the Sahel region of Africa, for use in other areas. The method uses data from the AVHRR instrument on board the NOAA series of polar-orbiting weather satellites. The AVHRR is capable of measuring spectral reflectance in five discrete bands. The AVHRR data are available in three spatial resolutions from the NOAA/NESDIS Satellite Data Services Division. These are (1) Local Area Coverage (LAC) at 1.1 km x 1.1 km resolution and available for only about 10% of each satellite orbit, (2) Global Area Coverage (GAC) at approximately 4 km resolution, and (3) the Global Vegetation Index (GVI) product at approximately 16 km (Kidwell, 1990, 1991).

The Normalized Difference Vegetation Index

Stressed vegetation has a higher reflectance than healthy vegetation in the visible (0.4 to 0.7 μm) region and a lower reflectance in the near infrared (0.7 to 1.1 μm) region of the electromagnetic spectrum. Vegetation indices take advantage of this differential response in the visible and near infrared (IR) regions of the spectrum. Perry and Lautenschlager (1984) provide an extensive review of vegetation indices based on LANDSAT and NOAA satellite data.

The Normalized Difference Vegetation Index (NDVI) uses NOAA/AVHRR spectral reflectance data in the visible (CH1: 0.58 to 0.68 μm) and near IR (CH2: 0.73 to 1.10 μm) channels. It is defined as follows:

$$NDVI = (CH2-CH1)/(CH2+CH1)$$

where CH1 and CH2 are reflectance, albedo, or radiance values from NOAA/AVHRR channels 1 and 2 (Gutman, 1987 and 1991). Water, clouds, and snow have higher reflectance in the visible region (CH1), and consequently the index assumes negative values for these features. Bare soil and rocks exhibit similar reflectance in both the visible and near IR regions, and the index values are near zero. The NDVI values for vegetation generally range from 0.1 to 0.6, the higher index values being associated with greater green leaf area and biomass (Tucker, 1979).

Time series of NDVI permit monitoring of the dynamic nature of vegetation phenology. Variations, or noise, in radiometric data are caused by instrument calibration, scan angle, sun angle, and atmospheric conditions. The effects of noise can be reduced in several ways: (1) compositing via the greenest pixel method, (2) using vegetation indices to compensate for variations of instrument calibration and cloud contamination, (3) aggregating and averaging the channel values over a defined geographic area and using these averages to calculate the vegetation index, and (4) smoothing the profile with curve fitting or filtering techniques (van Dijk et al., 1987). A weekly compositing period is generally used by NOAA for monitoring vegetation and pasture conditions. Malingreau (1986) used a three-week compositing period for monitoring crops such as paddy in Southeast Asia.

NOAA uses a blend of custom-designed and commercially available software to enhance, display, and analyze the AVHRR data. NOAA has developed the menu-driven MICVEGI software to process and display NDVI time series data. The NDVI data files are analyzed, both for vegetative trend through time and for change from a reference period through the extraction of time series profiles (Achutuni et al., 1991a).

An example of an NDVI time series plot for the Khon Khaen region of northeastern Thailand is shown in Figure 1. The bimodal nature of the curves is due to multiple cropping. Crops such as maize are planted in early April, whereas paddy or groundnut (peanuts) are cultivated later. An examination of Figure 1 shows that the index values during 1986 and 1988 were lower than those during March-April 1987, but they recovered as the season progressed. The time series for 1988 reveals the low index values associated with the drought recorded during the September-October period.

A commercial image processing system is used by NOAA to process the individual channel data files and to display the satellite imagery. Cloud screening is accomplished using temperature thresholds in the AVHRR thermal channel (CH4). Images are displayed using a color coordinate system based on hue, intensity, and saturation. The resulting color image can then be interpreted for qualitative information on vegetation conditions. The system is capable of image enhancement and classification operations. One can also display two saved images, from different crop seasons or years, side by side for visual comparison. Additionally, one can create a difference image to isolate areas of change between any two images. Color hard copies of the satellite imagery can be obtained using relatively inexpensive printers (Achutuni et al., 1991b).

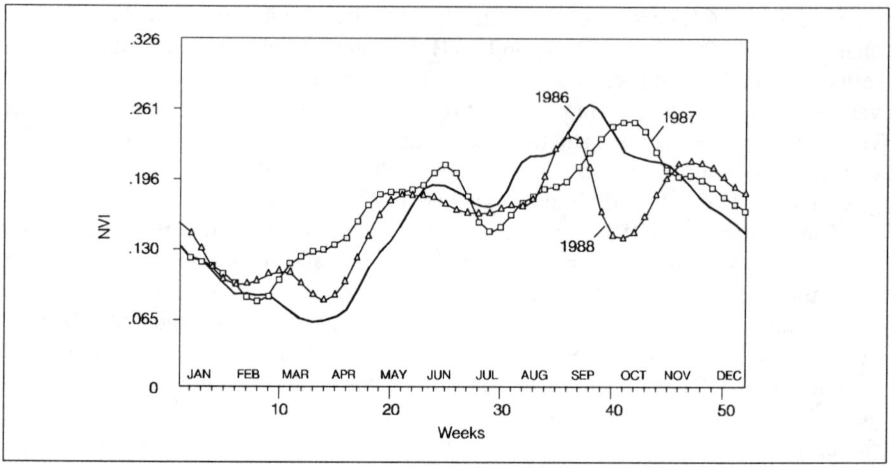

Figure 1. Normalized difference vegetation index time series for the Khon Khaen region of northeast Thailand.

SOUTHEAST ASIA: SATELLITE CROP-MONITORING EXPERIENCE

The Climate Applications Branch (CAB) of NOAA, National Environmental Satellite, Data, and Information Service (NOAA/NESDIS), and its affiliated Cooperative Institute for Applied Meteorology (CIAM) of the University of Missouri-Columbia designed and coordinated the Satellite Crop Monitoring Project in Southeast Asia. The overall objective of the project was to provide microcomputer-based satellite crop-monitoring hardware and software and associated training to the four participating countries of Indonesia, Malaysia, the Philippines, and Thailand. This project was Phase II of the Drought Early Warning Program, begun in 1983 to establish a crop monitoring tool for decision makers responsible for national food security in South and Southeast Asia.

NOAA/CAB/CIAM adapted satellite crop-monitoring technology, developed for monitoring large-area crop conditions in the Sahel region of Africa, for use in South and Southeast Asia. The method uses data from the AVHRR, capable of measuring spectral reflectance in five bands, on board the NOAA series of polar-orbiting weather satellites. The AVHRR data is very affordable and offers twice-daily temporal coverage, both important features in an operational assessment process where data availability and cost are primary considerations. In addition, these satellites provide reduced resolution data adaptable to microcomputer processing and analysis. The GVI resolution data were chosen for the Phase II satellite data analysis because they represent the most economical data source with the least quantity of data needed for satisfactory analysis results. Each country received a four-year (1985-89) historical data set of GVI resolution data for use in the workshop component of the project. Worldwide coverage GVI data files were subsetted to provide a southeast Asian window, and the training participants received weekly composited channel (channels 1, 2, and 4) and NDVI data files.

The NDVI data files are analyzed both for vegetative trend through time and for change from a reference period through the extraction of time series profiles. The channel data files are used to create satellite imagery, with channel 4 used for cloud identification. The resulting image can then be interpreted for qualitative information on vegetative conditions.

During the workshop, trainees were given an opportunity to become acquainted with all aspects of satellite data processing, image analysis, NDVI time series analysis, crop yield modeling, and preparation of the various assessment products. The assessment exercises were selected from documented incidence of drought in each of the participating countries. The training participants then tested the technology in their respective countries during the 1989-90 growing seasons and incorporated the results in their countries' monthly agroclimatic crop condition assessment bulletins. The GVI resolution AVHRR data files for this time period were forwarded to each country via diplomatic pouch for use during the test and evaluation phase.

Individual country presentations on the results of the test and evaluation segment of Phase II suggest that the satellite crop-monitoring technology is very useful for qualitative assessment of crop conditions, despite the obvious limitations of the coarser resolution GVI data. The intent was to introduce satellite crop-monitoring technology to these countries so that they could easily make the transition to NOAA's higher resolution AVHRR Level 1b LAC or GAC data as they become available to their respective countries. Most of the participating countries were able to fully use the technology and provide inputs to decision makers at the highest levels in government. This exchange of information and products between the technical assessors and the policy makers was one of the primary goals of the Satellite Crop Monitoring Project.

To build on the successfully implemented technology, the United Nations Economic and Social Commission for Asia and the Pacific (UN/ESCAP) and NOAA/NESDIS hope to expand the Early Warning Program to address most of the natural hazards faced by countries in the region of South and Southeast Asia by further expanding remote sensing concepts to natural resource management. Plans for Phase III specifically address drought monitoring, flood forecasting, cyclone/typhoon warning, teleconnections, natural resource management, and regional food security and policy issues.

INDIA: A NATIONAL PROGRAM INCORPORATING REMOTE SENSING

Background

Drought continues to be a factor of uncertainty in Indian agriculture despite significant scientific and technological developments, although the adverse impact has been minimized in recent years. Suitable monitoring mechanisms have evolved in the country to provide information necessary for tackling this hazard. Information on rainfall, aridity anomaly, and agricultural conditions are being collected and analyzed periodically. To improve timeliness, adequacy, and reliability, the use of satellite data is being investigated for possible integration into the existing drought surveillance system. The focus is on the assessment of agricultural drought conditions in terms of prevalence, relative severity level, and persistence through the season. Currently, the project involves 10 agriculturally important and drought-vulnerable states covering most of the country, and drought monitoring is restricted to the kharif (monsoon) season, which is agriculturally more important and is rainfall dependant. The National Agricultural Drought Assessment and Monitoring System (NADAMS) is sponsored by the Indian Department of Agriculture and Cooperation and is executed by the National Remote Sensing Agency (NRSA) in the Department of Space, Government of India (Thiruvengadachari, 1990).

Satellite Data and Methodology

NOAA's AVHRR sensor LAC data acquired daily at NRSA's earth station near Hyderabad are used to generate biweekly time-composited vegetation index (NDVI) imagery and data sets of India, with a 1 km pixel size. Experience has shown that time compositing of one week, as suggested by Holben (1986), resulted in unacceptable levels of residual cloud cover while periods longer than biweekly were inappropriate to monitor agricultural dynamics. Such data sets have been generated for May to December from 1986 to 1989 and throughout the year since May 1990. Comparison of NDVI temporal profiles of reporting year and the normal agricultural year provides an assessment of drought impact.

Extensive ground data (Figure 2) were collected for both methodology development and validation. Ground-truth campaigns were conducted with the cooperation of field officers of state government departments throughout the season to observe and report, in near-real time, agricultural information in specially designed standard formats. Calibration of NDVI data sets with ground data helped establish that even when NDVI data are averaged over the district (administrative unit), it is reflective of general vegetation dynamics and conditions and is driven by the rainfall cycle during the kharif season. The condition of specific crops, although not always directly detectable on the

District/Tahsil (1987/76/88/89)		Sample Site (1987/88/89)
Seasonal	**Weekly**	**Weekly**
Land use Cropped area Irrigation support Crop calendar Yield Rainfall Drought incidence	Rainfall Sown area Crop stage Crop condition Moisture stress Pest/disease attack	Geographical location Observed period Sky condition Rainfall Soil type/condition Sample site details Crop condition/yield
Eleven standard forms are used for reporting ground data to NRSA.		

Figure 2. Ground data collection on agricultural parameters.

NDVI profile, can be inferred with increasing reliability when the crop is rainfed and covers a large area uniformly distributed over the district. The NDVI is also seen to have a two- to four-week lag with a rainfall event. Comparisons with aridity anomaly reports and rainfall reports of the India Meterological Department showed encouraging correlation.

Biweekly drought bulletins were issued during June to December of 1989, 1990, and 1991 for 246 districts of 10 states and were sent to central and state government agencies, including district-level officers. The bulletin consists of three sections. The first section contains color-coded NDVI imageries of the state with district boundaries superimposed. The second section compares district-level greenness between the current reporting year and the previous two years, and the third section consists of a district-level assessment of drought prevalence and persistence and the relative severity level, based on the extent of NDVI anomaly in reference to a normal year. Since the printed bulletin takes about 10 days to be sent, preliminary alerts were sent by telex or telegram to drought-affected districts and concerned state government departments within 3-4 days of each biweekly period.

In response to the request of the relief commissioner of Andhra Pradesh State, subdistrict-level drought assessment was provided during 1991 in selected drought-affected districts of the state. Similar subdistrict-level drought monitoring was also attempted in Karnataka State during 1991. Detailed drought assessment studies for specific areas were carried out using LANDSAT Thematic Mapper (TM) and Indian Remote Sensing Satellite (IRS) data, which have significantly finer spatial resolution compared to NOAA's AVHRR data.

User Feedback

The response to a questionnaire distributed to the users of the biweekly drought bulletins indicated that the bulletins were regularly received (86%) and were timely (89%). It also brought out the need for better user awareness of the scientific basis for the bulletins. Users suggested larger-scale state NDVI imagery (1:1 million scale). The fact that users were willing to subscribe (35%) to the bulletin if it involved a cost is an encouraging sign of its usefulness. During discussion with the Indian Department of Agriculture and Cooperation, a directive was given to provide detailed drought assessment in terms of spatial variability and impact on crop production within the district in the future. The methodological improvements will include monitoring target areas within the district (such as kharif crop lands, major crop zones, and irrigated command areas versus rainfed areas) as well as developing models relating seasonal NDVI profile to crop yields. Instead of the biweekly bulletin, seasonal assessments starting after the peak crop season, updated monthly thereafter, will be provided.

Future Plans

Plans are under way to use ERS-1 Synthetic Aperture Radar (SAR) in Andhra Pradesh State for evaluating operational capabilities for drought stress detection during the cloud-covered kharif season. An airborne C-band SAR survey is also planned during the southwest monsoon season of 1992 to develop stress detection algorithms. The use of the thermal channel in LANDSAT Thematic Mapper will also be explored. Since drought monitoring and consequent relief management has sensitive political and financial implications, efforts are continuously made to ensure user acceptance of the satellite assessment of agricultural drought conditions before its integration into existing drought monitoring mechanisms.

UNITED STATES: AN INCREASED UNDERSTANDING OF DROUGHT THROUGH REMOTE SENSING

A New Method for Assessing Drought

In the past decade, a considerable number of climate and land surface products have been generated using NOAA polar-orbiting satellite measurements. Among land surface products, the vegetation index has been shown to be one of the most useful for description of land cover, vegetation classification, or vegetation phenology (Tucker et al., 1982; Tarpley et al., 1984; Justice et al., 1985). Unfortunately, the results were only marginal when attempts were made to use the vegetation index for estimating weather impact on vegetation.

Recently, a new algorithm has been developed at NOAA/NESDIS that has opened a different perspective on monitoring climate and land surface parameters and phenomena. The new method has shown particularly good results in drought detection and tracing, and also in estimating drought impact on vegetation and specifically on agricultural crops (Kogan, 1987 and 1990a). This method is based on the newly developed Vegetation Condition Index (VCI). The VCI was developed from the Global Vegetation Index product routinely produced by NOAA/NESDIS (Tarpley et. al., 1984). Of the data that this product contains, the NDVI was selected. Since NDVI has considerable noise that exceeds the environmental signal, a compound median filter was applied to the NDVI time series to remove this noise.

Further refinements have been made to separate the short-term weather signal in the NDVI data from the long-term ecological signal. This was done by normalizing smoothed weekly NDVI values relative to the amplitude of their change in each location during 1985-90. The weekly weather signal was amplified by ranking it on a linear scale with a minimum value of 0 and a maximum value of 100 in the six-year data set. The following expression defines VCI (Kogan, 1987 and 1990b).

$$VCI = \frac{(NDVI - NDVI_{min}) \cdot 100}{NDVI_{max} - MDVI_{min}}$$

where NDVI, $NDVI_{max}$, and $NDVI_{min}$ are the smoothed weekly Normalized Difference Vegetation Index, its multiyear maximum, and its multiyear minimum, respectively.

The VCI provides a numerical approximation of weather impacts on vegetation and also measures vegetation conditions. Low values of VCI indicate poor vegetation conditions and possibly unfavorable weather impacts; high values describe the opposite situation. For example, an area in central Illinois in the middle of the growing season will have NDVI values between 0.35 (minimum) and 0.52 (maximum). In central Kansas, minimum and maximum NDVI values are typically between 0.20 and 0.40. However, VCI values for maximum and minimum NDVI in both locations will remain the same, 0 and 100, characterizing unfavorable and favorable weather impacts, respectively.

Application of VCI for Tracking Drought

Of all U.S. droughts between 1985 and 1990, the 1988 drought was unique in its abnormal intensity, early season occurrence, enormous expansion, and considerable damage, especially to agriculture. Yield of grain crops decreased by 30%.

As the VCI showed, this drought was observed very accurately. It started in spring 1988 (Kogan, 1990b). At the beginning, this dryness covered only the northern Great

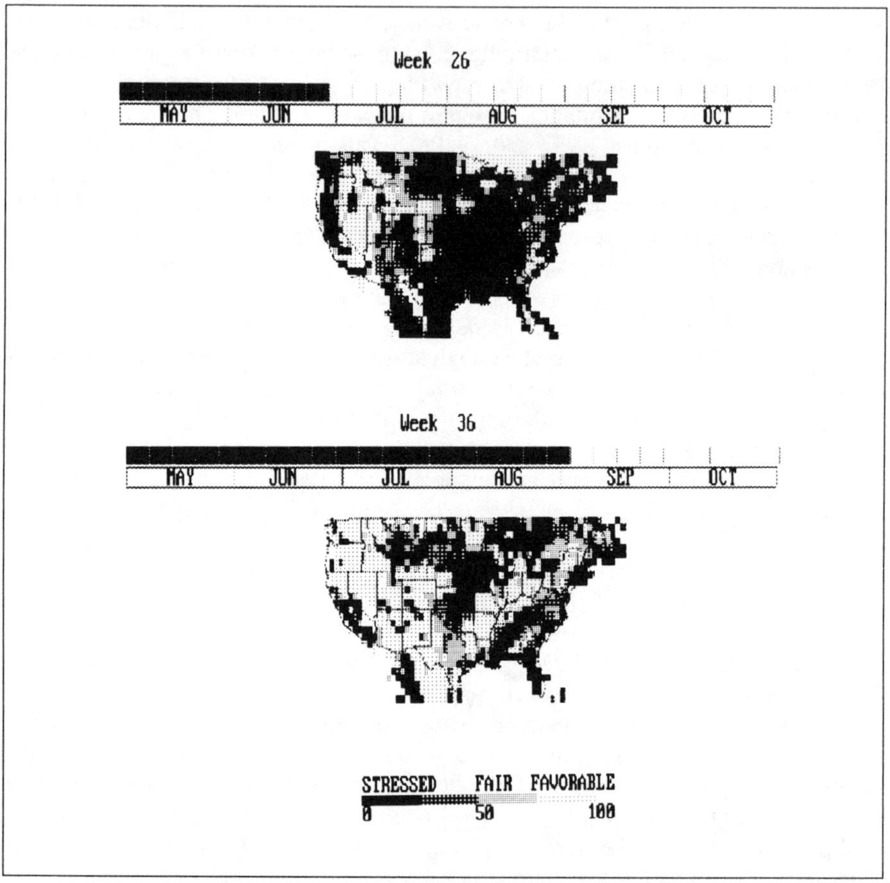

Figure 3. Vegetation Condition Index (a) at the end of June and (b) at the end of August 1988 in the United States.

Plains, the east, some western states, and southern Texas. By the end of June, it had expanded considerably. These observations were supported by ground measurements. As seen in Figure 3a, the drought area outlined by the VCI accurately portrays the precipitation deficit experienced during April-June and the resulting corn yield reduction (Figures 4a and 4c). Figure 3b shows the end of the late summer drought after July-August precipitation arrived. The VCI and precipitation-derived drought patterns at that time (Figure 4b) have a very good spatial correspondence.

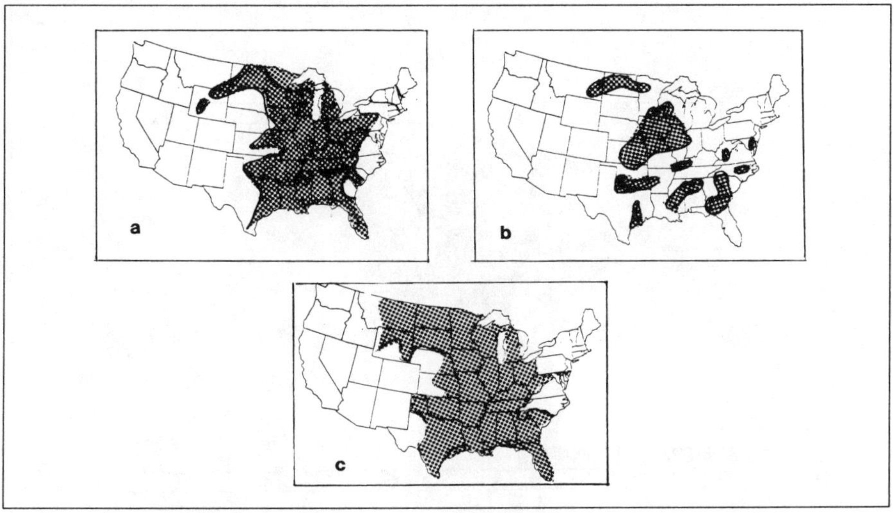

Figure 4. (a) Precipitation deficit of 3-6 in. during April-June 1988; (b) precipitation deficit of 2-4 in. during July-August 1988; and (c) reduction of corn yield relative to the 1985-88 mean for the United States.

The VCI provides accurate information not only for the cases with well-defined, prolonged, widespread, and intensive droughts, but also for localized, short-term, and poorly defined droughts. For example, a small-scale 1989 drought in the United States first developed in the central Great Plains. The pattern of this drought (Figure 5) matched the area of reduction of winter wheat yield in Kansas, and parts of Nebraska, Oklahoma, and Texas (Figure 6a). Later in the season, this drought damaged spring wheat and corn in North and South Dakota (Figure 6b). The area of this damage was in agreement with the VCI-derived drought pattern (Kogan, 1991).

Increased Understanding of Droughts

Nearly 1,200 first-order meteorological stations in the United States provide real-time weather information several times per day. This information is used for environmental monitoring purposes, including drought watch. However, the density of these stations is not sufficient to characterize regional drought. For example, in South Dakota, the ratio of weather stations to acres of land is 1:2 million. The weather data quite often are not available in real time. Moreover, weather parameters are physical in nature and are not sufficient to characterize agricultural drought and its impact on vegetation.

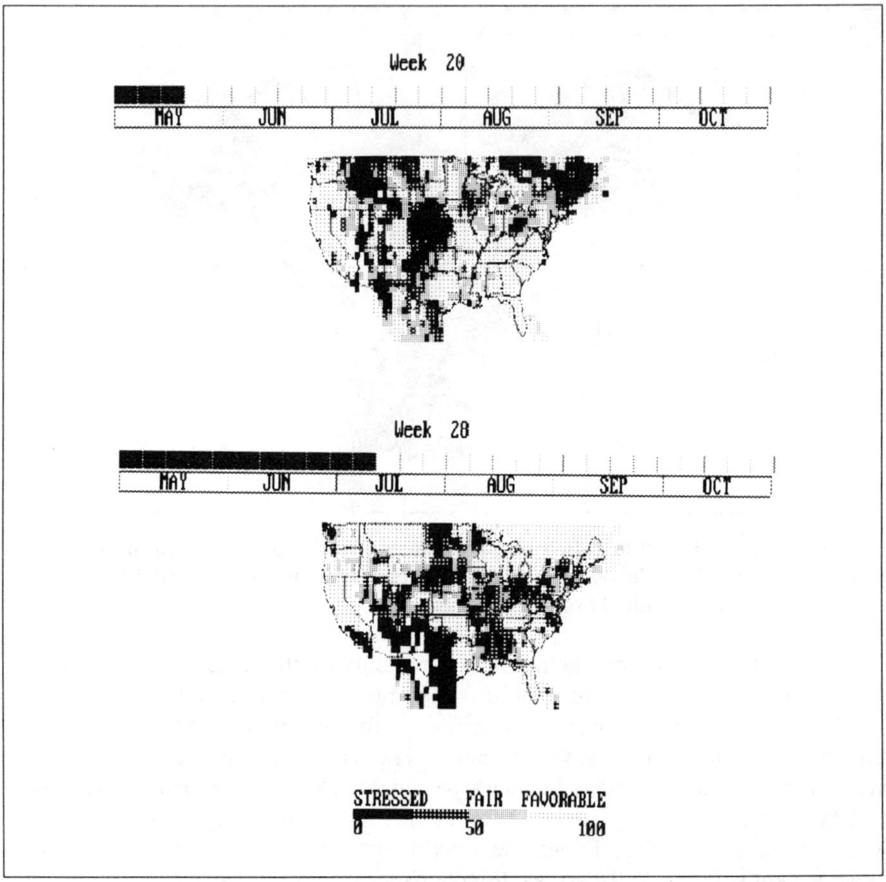

Figure 5. Vegetation Condition Index (a) in mid-May 1989 and (b) in mid-July 1989 in the United States.

Observations from satellite systems provide more timely information and much better spatial coverage. Moreover, AVHRR-derived vegetation indices reflect vegetation condition better than any other weather or climate variables. Therefore, using vegetation index data will increase our understanding of droughts. In addition to drought monitoring, we can address such problems as the calculation of soil moisture, evapotranspiration, prediction of yield, estimation of vegetation conditions and weather impacts, and stratification of ecological resources. These are areas in which some positive results have been obtained to date (Doraiswamy et al., 1991; Kunkel et al., 1991; Rasmussen, 1991) .

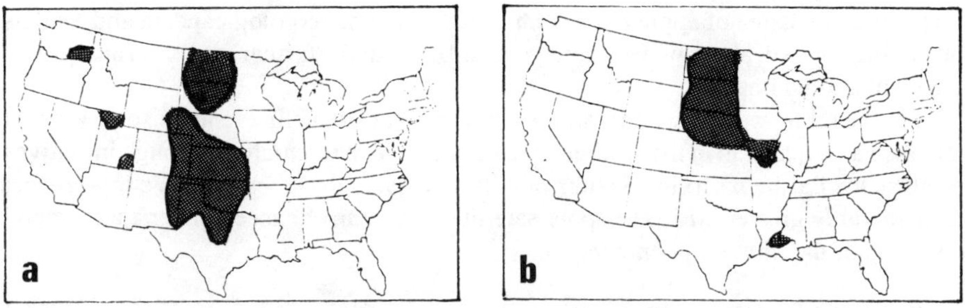

Figure 6. Reduction in (a) wheat yield and (b) corn yield relative to the 1985-88 mean in the United States.

Satellite-Based Drought-Watch System

Almost every year, millions of acres of agricultural land in the world are affected by droughts. The coincidence of droughts in several principal agricultural areas has always had very unfavorable effects, quite often leading to devastating economic and social impacts. As mankind faces the inevitable recurrence of drought in the future, the development of drought-mitigation measures is an important task. The first step in this endeavor is a drought-watch capability. A satellite-based drought-watch system will serve this goal.

Presently, the development of such a system, based on available satellite data, has potential. Weekly vegetation index data of 16 km resolution are available from NOAA/NESDIS in real time for any part of the globe between 75°N and 55°S latitude. The VCI, derived from these data, has the ability to detect drought and measure its intensity, duration, dynamics, and impacts on vegetation. This VCI can serve the practical purpose of drought monitoring on continental and regional scales. For smaller geographic areas and site-specific studies, finer resolution vegetation index data can be used in a similar manner.

SUMMARY AND CONCLUSIONS

As noted throughout this chapter, satellite remote sensing, and particularly the use of data from the AVHRR sensor onboard NOAA's polar-orbiting satellites, is expanding rapidly in the area of drought monitoring and assessment. It would be misleading, however, if the reader were to infer (1) that all work related to this area of applied research had been discussed in this brief chapter or (2) that satellite-based remote sensing can serve as a stand-alone tool for drought assessment and monitoring. Other researchers, many of whom are cited in the reference section of this chapter, are actively expanding this area of applied research. Traditional meteorological data and models are being applied, along with remotely sensed data, in the further study and analysis of drought and its impacts.

In the future we will witness an even greater array of satellites and sensors. We will also see an expansion in the volume of data available to researchers through initiatives such as the Earth Observing System and its associated data archives. We may expect considerably greater efforts to apply satellite data to an enhanced understanding and analysis of the drought phenomenon.

ACKNOWLEDGMENTS

Appreciation is extended to the National Oceanic and Atmospheric Administration, National Environmental Satellite, Data and Information Service, Office of Research and Applications, Satellite Applications Laboratory (with which Dr. Felix Kogan is affiliated), and Satellite Research Laboratory (with which Dr. V. Rao Achutuni is affiliated). Appreciation is also extended to the National Remote Sensing Agency of India, with which S. Thiruvengaddachari is affiliated. Grateful thanks are due to the Secretary, Department of Space, Government of India; Director, National Remote Sensing Agency; and Secretary, Department of Agriculture and Co-operation, Government of India, for the valuable support received during the first phase of the development of NADAMS. The contribution from various colleagues in NRSA and other centers of the Indian Space Research Organization is also gratefully acknowledged.

ABOUT THE AUTHORS

Gary E. Johnson is the deputy manager of the Earth Resource Observation Systems (EROS) Data Center Project for Hughes STX Corporation. (Hughes STX provides scientific and systems expertise to the Data Center in meeting its diverse mission requirements.) A geographer with more than 25 years of experience, Dr. Johnson has worked for several universities, state and fderal governments, and the United Nations Environment Programme. He has worked and traveled extensively in Africa and Southeast Asia. He is a member of several professional societies and has published extensively in the area of applied remote sensing.

V. Rao Achutuni is a meteorologist with the National Oceanic and Atmospheric Administration, National Environmental Satellite, Data, and Information Service (NOAA/NESDIS), Satellite Applications Laboratory, Physical Sciences Branch (SAL/PSB) in Camp Springs, Maryland. After receiving a Ph.D. from the University of Oklahoma, Dr. Achutuni joined the Cooperative Institute for Applied Meteorology (CIAM) at the University of Missouri in Columbia, Missouri, where he continues to serve as an adjunct associate professor in the Department of Atmospheric Science. His research interests are the design and implementation of global early warning and food security systems, satellite applications to agriculture and natural disaster management, crop yield modeling, applied statistics, and international training and technology transfer. He is a member of the American Meteorological Society and Sigma Xi.

S. Thiruvengadachari graduated in hydrology and water resources after basic qualification in civil engineering. He is the head of the Water Resources Division of the National Remote Sensing Agency in Hyderabad, India, where he guides remote sensing applications in water resources management. His major interests are irrigation management, drought monitoring, flood mapping, and management and water quality surveys. Mr. Thiruvengadachari is the project manager of the National Agricultural Drought Assessment and Monitoring, developed at the Department of Space, Government of India. He has more than seventy publications to his credit.

Felix Kogan received his Ph.D. in agricultural meteorology. He is a physical scientist at the Satellite Research Laboratory of the National Environmental Satellite Data and Information Service, NOAA. During his career, he has studied climate and its impacts on the environment, climate change and its consequences for the biosphere, and modeling of environmental impacts on agriculture. Dr. Kogan's research interests center on using satellite data for interpretation of land surface features and diagnostics of weather- and climate-driven changes in the environment. He is the author of more than 50 publications.

REFERENCES

Achutuni, V. R.; S. Weisman; T. Phillips; and R. Terry. 1991a. Large-scale vegetation monitoring in Southeast Asia. *20th Conference on Agricultural and Forest Meteorology*. American Meteorological Society, Boston.

Achutuni, R.; R. Terry; and S. Weisman. 1991b. *Satellite Crop Monitoring Project (Phase II). Final Report to the United Nations Economic and Social Commission for Asia and the Pacific (UN/ESCAP), Bangkok, Thailand.* Available from NOAA/NESDIS Physical Sciences Branch, Rm. 600, World Weather Building, 5200 Auth Road, Washington, D.C. 20233.

Doraiswamy, P. C.; and E. Perry. 1991. The relationship between satellite-based surface temperature and vegetation indices over semi-arid rangeland watershed. *Proceedings of the 10th Conference on Biometeorology and Aerobiology*. American Meteorological Society, Boston.

Gutman, G. G. 1991. Vegetation indices from AVHRR: An update and future prospects. *Remote Sensing of the Environment* 35:121-36.

Gutman, G. G. 1987. The derivation of vegetation indices from AVHRR data. *International Journal of Remote Sensing* 8:1235-42.

Holben, B. N. 1986. Characteristics of maximum-value composite images from temporal AVHRR data. *International Journal of Remote Sensing* 7(11):1417-34.

Justice, C. O.; J. R. G. Townshend; B. N. Holben; and C. J. Tucker. 1985. Analysis of the phenology of global vegetation using meteorological satellite data. *International Journal of Remote Sensing* 6(8):1271-1318.

Kidwell, K. B. 1991. *NOAA Polar Orbiter Data User's Guide: TIROS-n, NOAA-6, NOAA-7, NOAA-8, NOAA-9, NOAA-10, NOAA-11, and NOAA-12.* Published by the U.S. Department of Commerce, NOAA/NESDIS Satellite Data Services Division, Washington, D.C. 20233.

Kidwell, K. B. 1990. *Global Vegetation Index User's Guide.* Published by the U.S. Department of Commerce, NOAA/NESDIS Satellite Data Services Division, Washington, D.C. 20233.

Kogan, F. N. 1991. Drought Watch System Using Satellite Observations. *Proceedings 7th International Conference on Interactive Information and Processing Systems for Meteorology, Oceanography, and Hydrology*; pp. 379-82. American Meteorological Society, Boston.

Kogan, F. N. 1990a. Remote sensing of weather impacts on vegetation in non-homogeneous areas. *International Journal of Remote Sensing* 11(8):1405-19.

Kogan, F. N. 1990b. Monitoring the 1988 U.S. drought from satellite. *Proceedings 5th Conference on Satellite Meteorology and Oceanography*; pp. 186-89. 3-7 September, London.

Kogan, F. N. 1987. Vegetation index for areal analysis of crop conditions. *Proceedings of the 18th Conference on Agricultural and Forest Meteorology*; p. 103. American Meteorological Society, Boston.

Kunkel, K. E.; S. E.Hollinger; F. N. Kogan. 1991. Soil moisture/evaporation/precipitation feedback: A case study of the 1988 drought. *Proceedings of the 10th Conference on Biometeorology and Aerobiology.* American Meteorological Society, Boston.

Malingreau, J. P. 1986. Global vegetation dynamics: Satellite observations over Asia. *International Journal of Remote Sensing* 7:1121-46.

Perry, C. R.; and L. F. Lautenschlager. 1984. Functional equivalence of spectral vegetation indices. *Remote Sensing of the Environment* 14:169-82.

Rasmussen, M. S. 1991. Crop yield assessment in Senegal and Burkina Faso using NOAA AVHRR data. *Proceedings 5th AVHRR Data User's Meeting*; pp. 421-26. Tromso, Norway, 25-28 June.

Tarpley, J. P.; S. R. Schnieder; R. L. Money. 1984. Global vegetation indices from NOAA-7 Meteorological satellite. *Journal of Climate and Applied Meteorology* 23:491.

Tucker, C. J.; J. Gatlin; S. R. Schneider; M. A. Kuchinos. 1982. Monitoring large scale vegetation dynamics in the Nile delta and river valley from NOAA AVHRR data.

Proceedings of Conference on Remote Sensing of Arid and Semi-Arid Lands; p. 973. Cairo, Egypt.

Tucker, C. J. 1979. Red and photographic infrared linear combinations for monitoring vegetation. *Remote Sensing of the Environment* 8:127-50.

Thiruvengadachari, S. 1990. Satellite surveillance for improved countrywide monitoring of agricultural drought conditions. *Proceedings of the National Symposium of Remote Sensing for Agricultural Applications*; pp. 389-407. New Delhi.

van Dijk, A.; S. L. Callis; C. M. Sakamoto; and W. L. Decker. 1987. Smoothing vegetation index profiles: An alternative method for reducing radiometric disturbance in NOAA/AVHRR data. *Photogrammetry Engineering and Remote Sensing* 53:1059-67.

Chapter 4
Assessing the Regional Consequences of Drought: Putting the MINK Methodology to Work on Today's Problems

William E. Easterling

INTRODUCTION

How much confidence can be placed on past assessments of the impacts of drought on a region's resource base and associated economy? Beyond crude, highly aggregated estimates, what can be said about the many biophysical and economic impacts of the 1988 drought in the United States? What would be the consequences of a string of droughts such as a modern recurrence of the notoriously hot and dry conditions of the 1930s Dust Bowl climate that affected large portions of the United States? The answers to these questions are not easy to obtain. The problem, in large measure, is that drought impact methodologies are poorly developed to undertake the complex task of estimating the consequences of drought (see Chapter 1 for a comprehensive list of economic, social, and environmental impacts). The call for improved drought impact methodologies comes from a long line of research (for example, Yevjevich et al., 1978; Rosenberg, 1980; Wilhite and Easterling, 1987; Riebsame et al., 1990). Riebsame et al. (1990) conclude that much has been learned in the last decade about the interplay between climate and natural resources systems. Simulation models are widely available in almost all drought-sensitive sectors, but they have yet to be integrated into a single program of impact analysis aimed at assessing the overall impact of drought on the entire natural resource base of a region. The purpose of this chapter is to describe a methodology that was developed to assess the overall response of a region's natural resource base and associated economy to long-term changes in climate and to demonstrate the applicability of this methodology to the assessment of drought impacts.

Background

In late 1988, the U.S. Department of Energy commissioned Resources for the Future, Inc. (RFF), to develop a methodology for estimating the regional impacts of and responses to climate change (Rosenberg and Crosson, 1991). At the time, previous climate impact methods suffered from one or more deficiencies that undermined their relevance to policy making. Deficiencies specific to drought impact assessment include the following: (1) the full range of available technologies, management techniques, and policy tools for adjusting to climate variation were not fully considered in the calculus of impacts; and (2) the focus had been almost exclusively on a single economic sector such as agriculture with no consideration of important linkages to other sectors such as water and energy.

The methodology described here was developed, in part, to improve the modeling of adjustments to climate fluctuation and to fully consider intersectoral linkages in the estimation of climate impacts. A regional approach was used in which the central question asked was how climate variation affects all of a region's climate-sensitive resource sectors. The four-state region of Missouri-Iowa-Nebraska-Kansas (MINK, see Figure 1) was chosen to develop the methodology, although in principle the methodology is transportable to other regions. The MINK region was chosen because of its topographic and economic homogeneity and because its economy, relative to the rest of the country, is highly specialized in natural resource-based sectors that are sensitive to climate fluctuations such as drought.

An analytical framework was developed that first examined the current functioning of a regional-scale economy in the absence of drought. Such a framework provides a detailed baseline description of the region. Second, a scenario of drought was imposed

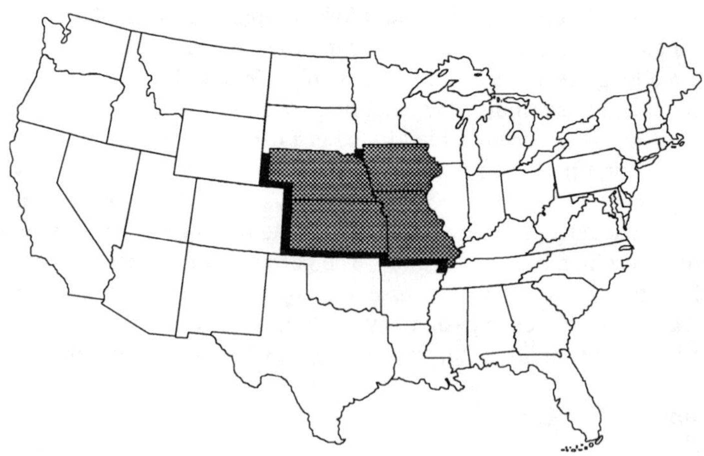

Figure 1. The four-state region (Missouri, Iowa, Nebraska, Kansas) chosen to develop the methodology.

on the region to determine how baseline resource productivity (e.g., streamflow, runoff, crop yields) is altered, in the absence of any adjustments by resource managers and associated institutions. Such a scenario is the equivalent of a short-term drought that gives people and institutions little time to respond. Some droughts persist long enough to allow adjustments to take place. Thus, the third part of the analytical framework was to impose a longer-term (ten-year) drought to determine how baseline resource productivity is altered after adjustments have occurred. The framework allows the examination of drought impacts at the farm/firm level and at the regional level.

The methodology was developed with full consideration of the important resource sectors in the MINK region (e.g., water, agriculture, forestry, and energy) and their linkages. Space limitations preclude description of the full suite of resources in this chapter. Instead, emphasis will be given to agriculture and water resources and their links to the rest of the MINK economy.

CURRENT CLIMATE AND SCENARIO OF DROUGHT

The MINK region's climate is typically continental. This leads to large seasonal swings in temperature and precipitation, with cold dry winters followed by hot summers where precipitation falls off in a gradient from east to west. Precipitation in the MINK region is determined by two physiographic features: the Rocky Mountains, which block the incursion of Pacific moisture from the west, and the Gulf of Mexico, which provides summer moisture from the southeast. Distance from the Gulf of Mexico and the rain shadow of the Rocky Mountains account for the distance decay of precipitation from east to west. Precipitation is least in winter and greatest in summer, except in Missouri, where the maximum occurs in the spring.

The scenarios of drought chosen for this study were drawn from the weather of the decade of the 1930s. This period, typically referred to as the Dust Bowl because of the devastating dust storms unleashed by extended periods of hot and dry conditions, is the hottest and driest on record. Differences in temperature and precipitation between the 1930s and the 1951-80 baseline period are shown in Table 1. In the 1930s, winters were warm in Missouri and Iowa, as were autumns in Nebraska and Kansas. Precipitation was, on average, lowest in spring and summer. Interannual variability in temperature was high in the 1930s while interannual variability in precipitation was low.

The droughts of the 1930s serve as a "worst case" benchmark against which to measure the current vulnerability of the MINK region to a severe and protracted drought. In the analyses that follow, the climate of the 1930s is compared with the climate of the period defined by the World Meteorological Organization convention as normal (1951-80).

	Winter(D,J,F)		Spring(M,A,M)		Summer(J,J,A)		Fall(S,O,N)	
	Temp. (°C)	Precip. (mm)	Temp. (°C)	Precip. (mm)	Temp. (°C)	Precip. (mm)	Temp. (°C)	Precip. (mm)
Missouri								
Mean	1.0	16	0.0	-23	1.1	-21	0.6	-1
SD	0.3	-3	0.0	18	0.3	7	0.2	12
Iowa								
Mean	1.1	6	0.3	-53	1.2	-28	0.6	16
SD	0.9	-6	-0.2	-4	0.1	11	0.2	20
Nebraska								
Mean	0.6	4	0.7	-23	1.6	-54	1.0	-21
SD	0.8	-2	0.1	7	-0.1	-13	-0.1	-13
Kansas								
Mean	0.9	0	0.6	-19	1.3	-59	1.0	-24
SD	0.4	-12	-0.1	-6	0.1	25	0.0	-1

Table 1. Differences between 1931-40 and 1951-80 in seasonal mean monthly temperature and mean seasonal precipitation totals and their standard deviations in the MINK states.

A BASELINE DESCRIPTION OF AGRICULTURAL AND WATER RESOURCES IN THE MINK REGION

In the full MINK study, a complete baseline description of the MINK economy was constructed. To illustrate the approach, that part of the baseline description that focuses on agriculture and water resources will be discussed here. Furthermore, in the crop-modeling work reported below, only those crops grown under both dryland and irrigated conditions in MINK—corn, sorghum, and wheat—are emphasized. For analysis of the overall impact of drought on the regional economy, information on soybeans is also provided.

Statistics to describe the MINK economy at the time the study began were only available through 1987, so the base period is 1984-87. In the base period, the MINK region produced 34% of the nation's value of corn for grain, 30% of its soybeans and winter wheat, and 50% of its grain sorghum. Mean yields of these four crops were higher and costs of production per unit volume were less in MINK than in the rest of the United States. Nebraska and Kansas combined to account for 94% of the irrigated land in the region. MINK also produced 28% of the nation's cattle and calves and 42% of its hogs.

Input-output modeling of the MINK economy using a modified version of the U.S. Forest Service IMPLAN model (discussed below) showed that a change of $1.00 in on-farm production generated an additional change of $0.65 in production in the rest of the regional economy. The regional impact of meat-packing, for example, was greater, with a change of $1.00 generating an additional change of $1.57 in the rest of the regional economy.

The MINK economy is much more specialized in agriculture than the nation as a whole. The ratio of farm income to total income in the MINK region is almost three and a half times the ratio for the nation as a whole. Nebraska is the most specialized in agriculture while Missouri is the least specialized. Although MINK is highly specialized in agriculture relative to the nation, manufacturing and services account for the largest share of regional income. This last observation can be somewhat misleading since much of MINK's manufacturing is in agriculturally related firms.

The MINK region encompasses parts of four major river basins: the Missouri, the Arkansas-White-Red, and the upper and lower Mississippi. Significant portions of these river basins fall outside of the MINK region, thus emphasizing the importance of climatic and other circumstances upstream and downstream from the region. Irrigation accounts for approximately half of the water withdrawals from ground and surface water sources, and it accounts for 89% of the consumptive use of water in MINK. Groundwater provides about 80% of all irrigation water and 48% of all water withdrawals in the region. The western MINK states are heavily reliant on groundwater from the Ogallala Aquifer for irrigation. The Ogallala Aquifer is a deep formation with very low recharge rates, particularly in the southern part of the region. Heavy pumping has mined groundwater in the Ogallala to the point that water tables have fallen precipitously in some parts of the MINK region. It is doubtful that current rates of groundwater mining in this region are sustainable. A severe or protracted drought would further stress groundwater resources in the western MINK.

Conflicts over the uses of water supplies in and upstream from the MINK region have arisen in recent years. On the mainstem of the Missouri River, for example, a growing demand to protect recreation and wildlife habitat upstream of the MINK region has come into direct conflict with water-borne transportation needs downstream. Recent droughts have further stressed such conflicts. Future droughts will similarly stress conflicts over alternative uses of water in the MINK region.

MODELING THE EFFECTS OF DROUGHT

Agricultural Analysis. Modeling the effects of drought on the MINK region begins at the farm level with the use of a semimechanistic crop model to predict the response of crop yields to drought. The Erosion Productivity Impacts Calculator (EPIC) was the model chosen for the MINK study. EPIC was developed by the U.S. Department of Agriculture to estimate the long-term effects of soil erosion on the nation's agricultural productivity (Williams et al., 1984). Modifications were made to EPIC to permit accurate simulation of crop response to drought. EPIC was modified as follows:

1. Nearly 50 representative farms were constructed to model the diversity of farming operations across the MINK region. Information on cultural practices and soils was gathered for each representative farm from interviews with agricultural experts at each of the four MINK land grant universities. Such information was used to construct input data sets for the EPIC model.

2. EPIC operates on a daily time step. Daily climate data were compiled for the 1930s and the 1951-80 baseline for 17 stations across the MINK region. These data include maximum and minimum temperature and precipitation. Daily relative humidity and solar radiation values were not available for the 1930s, so they were generated from monthly values with EPIC's stochastic weather generator.

A major reason for using EPIC is its flexibility, which allows simulation of alternative crop varieties, varying crop rotations, different input amounts, cultural practices, pest control, irrigation, and farm economics. This flexibility permits the modeling of adjustments that farmers will make in response to long-term drought.

EPIC was validated for realism in three ways. First, EPIC simulations of crop yield under current weather and technology were compared with the observed yields extracted from USDA. Second, the simulated yields were compared with judgments of experts who helped build the representative farms. Third, the simulations were compared with the range of yields reported from experimental results in the literature. The results of the simulations are reported fully in Rosenberg et al. (1992). In general, the EPIC simulations were found to be accurate enough to permit a realistic assessment of crop response in the region to drought.

Water Resources Analysis. EPIC was used to estimate the effects of drought on irrigation water requirements (Frederick, 1991). No model was readily available for estimating the effects of drought on streamflows. Ideally, natural flows in the 1930s would be compared with those of today. However, several alterations have been made to the MINK river systems (e.g., dams, diversions) since the 1930s, which distort direct comparisons between the two time periods. Consequently, only a small number of gauging stations (out of thousands) apparently have not been altered by human activity since the 1930s. Those stations can be used as a proxy for estimating natural streamflows in the MINK region.

Gross evaporation in the reservoirs of the MINK region was estimated using the Harbeck (1962) and Penman (1948) methods. Mean precipitation values for the 1930s and 1951-80 baseline were subtracted from gross evaporation to calculate net evaporation.

IMPACTS OF SHORT- AND LONG-DURATION DROUGHTS ON CROP YIELDS AND WATER DEMAND AND SUPPLY

EPIC does not directly produce reliable estimates of crop yields under short-term (one- to three-year) droughts since a slightly longer period is needed for its estimates to stabilize. Thus the weather of the entire decade of the 1930s was used to simulate crop yields for both short- and long-term droughts. To emulate a shorter-term drought, it was assumed that farmers and water managers do nothing to respond to the droughts other than to increase irrigation applications on irrigated lands since the short lead time and duration of the drought leave little opportunity for adjustments. Thus, the mean response of crop yields to the weather of the 1930s in the MINK region is assumed to be a composite of a short-term drought. In the long term (four-year droughts or longer), however, resource managers will adjust to drought conditions by altering operations to reduce climate risk. Hence, the simulation of resource productivity under the 1930s droughts with adjustments is a realistic appraisal of the impacts of a long-term drought.

Crop Yields. The EPIC simulations of yields were averaged by crop across all representative farms in the MINK region. Were farmers simply not to respond to a recurrence of the droughts of the 1930s, dryland corn and sorghum would yield 25% and 22% lower than today, respectively (Figure 2). Yield losses are primarily due to

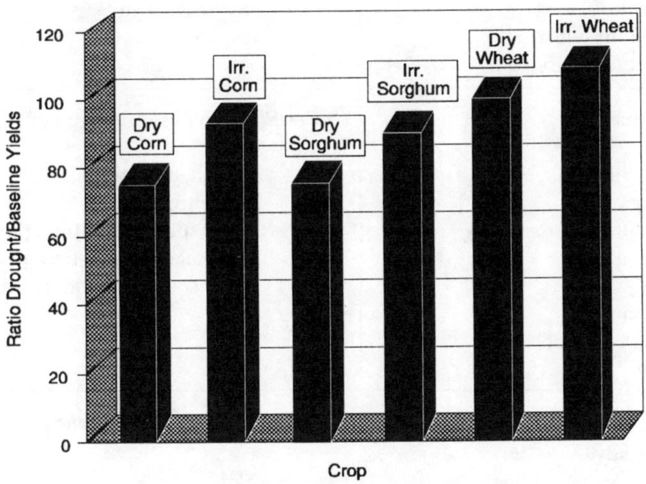

Figure 2. Yield response to short-term droughts in the MINK region: No farmer adjustments.

the speeded-up development of the crop caused by the warmer temperatures. The crops, in short, develop too rapidly in the warmer, drier climate to achieve adequate grainfill. The mean yield of dryland wheat is essentially no different from the baseline (Figure 2). Wheat is probably aided by the milder winters of the 1930s, allowing it to break from dormancy earlier. Loss of corn and sorghum yields from the drought conditions is mitigated by irrigation. Irrigated corn is reduced by 7% and irrigated sorghum by 10% (Figure 2). The yield of irrigated wheat increases by 8%.

Farmers will make adjustments to long-term droughts in order to minimize rising production costs. In the MINK study, a number of simple, low-cost adjustments were examined to determine their efficacy in offsetting drought-induced yield losses. Those adjustments are listed in Table 2. Analyses were conducted to determine which of the adjustments listed in Table 2 make a difference in the EPIC model's sensitivity to

Planting and Harvesting Practices	*Variety and Crop Selection*
Shift planting dates Plant a mix of varieties with different pollination times Lister plant **Plant seeds deeper in the soil profile** **Reduce plant densities** Reduce artificial drying	**Switch to shorter- or longer-season variety** **Switch to more stress-tolerant crop** **Switch marginal areas to pasture or range** Grow more specialty crops
Land Management	***Fertility and Pest Management***
Reduced tillage **Fallowing** **Furrow diking** Stubble mulching Ridge till Strip farming Elimination of double cropping	Use more herbicides and less cultivation **Reduce nitrogen applications** Incorporate nitrogen-fixing winter cover crop into rotation
Adjustments in bold type can be represented readily in the version of the EPIC model used for the MINK study.	

Table 2. Low-cost adjustment strategies identified by local agricultural experts for dealing with climate fluctuation.

climate variation. Those adjustments that prove effective in reducing drought sensitivity were then applied to the MINK region as a whole. The adjustments that proved effective included planting earlier in combination with switching to longer-season cultivars, thus permitting farmers to take advantage of the longer growing season and the use of furrow diking to harvest water when it rains.

When droughts last long enough for farmers to respond, their adjustments are most effective in offsetting yield losses for sorghum, in both its dryland and irrigated form (Figure 3). The adjustments improved sorghum yields to 86% of the baseline from 76% of the baseline without adjustment. Adjustments improve dryland corn yields only slightly and eliminate all losses in irrigated corn. Earlier planting and furrow diking do not apply to wheat. The longer-season varieties encounter temperatures too high to be beneficial to winter wheat. Thus, wheat was not as positively responsive to adjustments as the other crops. This is the result of the choice of adjustment strategies rather than an indication of the intrinsic adjustability of wheat.

Water Demand and Supply. In the EPIC simulations, irrigation water was supplied to crops based on physiological demand, irrespective of the cost or availability of irrigation water. Average changes in irrigation requirement caused by the 1930s drought conditions are shown in Table 3. In the case of short-term drought (absent adjustments), irrigation requirement increased by 11% in wheat to 24% in corn. In the

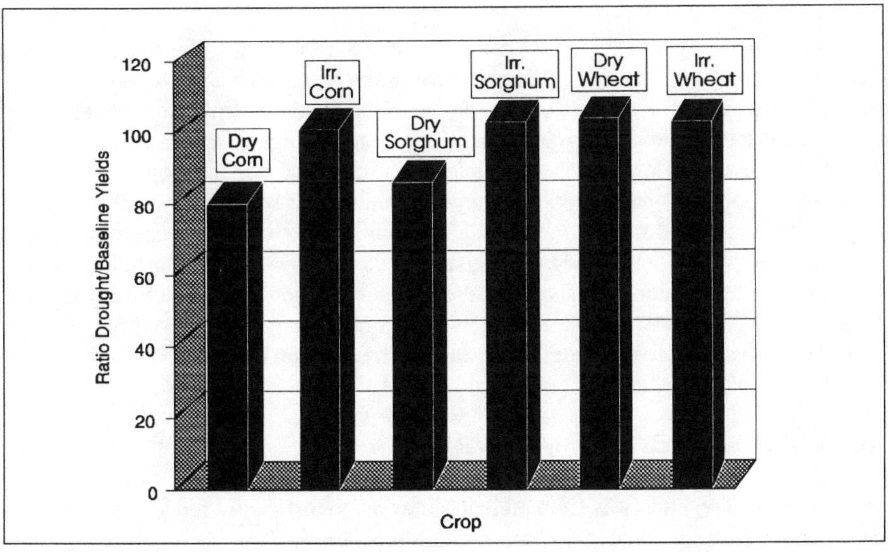

Figure 3. Yield response to long-term drought in the MINK region: With farmer adjustments.

	No Adjustments	With Adjustments
Corn	24	43
Sorghum	20	39
Wheat	11	4

Table 3. EPIC-simulated percentage changes in irrigation requirement under short-term drought (no adjustments) and short-term drought (with adjustments).

case of long-term drought, adjustments actually cause irrigation requirements to rise still higher because of the extra water needed for the longer-season varieties.

After scaling up the individual EPIC farm simulations according to the area they represent and assuming all demand can be met by current irrigation systems and water supplies, water withdrawals under a short-term drought like a composite of the 1930s drought conditions would increase by 39% in Nebraska and 14% in Kansas. On the supply side, proxy gauging stations from the 1930s indicate that streamflows were 28% lower than their 1951-80 average on the Missouri and Upper Mississippi basins and 7% lower on the Arkansas basin. According to estimates of evaporation from reservoirs, the 1930s drought conditions would increase evaporation enough to reduce streamflows by only an additional 1%.

Were the drought to persist for more than one to three years, it is virtually certain that the increased irrigation requirement would not be sustainable in Nebraska and Kansas. Competition for water coming into the MINK region from other states suggests that Kansas and Nebraska farmers are not likely to acquire additional surface water rights. Farmers are also not likely to be able to afford the additional energy costs of extra irrigation pumpage when the yields they will attain will be lower than the baseline yields. Instead, irrigators are likely to adapt to more scarce irrigation water supplies by investing in more efficient irrigation systems such as low-pressure sprinkler heads, switching to less water-intensive crops, and returning to dryland cultural practices (Kromm and White, 1986). Indeed, a recurrence of the drought conditions of the 1930s would likely accelerate the widespread decline in irrigation in western MINK that has already begun (Nellis, 1987).

ECONOMIC IMPACTS OF DROUGHT

The farm-level analysis showed that yields are significantly reduced for dryland corn and sorghum, and irrigation requirements are increased for all crops in both short- and long-term droughts. As a result, production costs in the MINK region would surely rise. Easterling et al. (1992) suggest that production would decline in the MINK region

	Without Adjustments		With Adjustments	
	Value (millions, 1982 $)	Change in 1984-87 Production (%)	Value (millions, 1982 $)	Change in 1984-87 Production (%)
Corn	-1,644	-21.3	-1,729	-22.4
Wheat	-14	-0.8	+361	+19.7
Sorghum	-215	-17.1	-35	-2.8
Soybeans	-789	-23.0	-542	-15.8
Total	-2,662	-18.7	-1,945	-13.7

Table 4. Changes in the value of crop production under short-term drought (without adjustments) and long-term drought (with adjustments).

because the baseline level of production would no longer be profitable with higher production costs.

For simplicity, it was assumed that production costs would increase under short-term drought conditions (no adjustments) such that the decline in production would be directly proportional to the decline in yields. In accounting for adjustments by farmers in response to long-term drought, production is tied not only to trends in yields but also to the conversion of some irrigated corn into dryland production of wheat and sorghum. Those adjustments would offset some of the rise in production costs in irrigated corn.

Table 4 shows changes from the 1984-87 baseline in the value of production of corn, wheat, sorghum, and soybeans under the 1930s drought conditions. Changes are shown for the short-term drought (without adjustments) and long-term drought (with adjustments) cases. The total value of production for all four crops declines by $2,662 million (-18.7%), based on 1982 prices under the short-term drought. For long-term drought, the adjustments offset a portion of the decline in value of production, but the MINK region still remains 13.7% below the baseline.

Effects on the Regional Economy. The specialization of the MINK region in agriculture suggests that major changes in the value of agricultural production caused by drought will ramify to other linked economic sectors in the region. Such ramifications can be represented quantitatively with the use of multipliers or technical coefficients derived in an input-output economic model. Input-output models track flows of goods and services between sectors and ultimately to final demand. A county-level input-output model of the entire United States, known as IMPLAN (U.S. Forest Service, 1989), was modified to focus only on the MINK region. The model is discussed more fully by Bowes and Crosson (1991). Here, only the results the model gives in

Export Scenario 1*				Export Scenario 2**			
No Adjustments		With Adjustments		No Adjustments		With Adjustments	
($mil.)	(% total prod.)	($mil.)	(% total prod.)	($mil.)	(% total prod.)	($mil.)	(% total prod.)
-4,123	1.3	-1,403	0.5	-29,881	9.7	-17,622	5.7

*Assumes feed grain decline falls on exports
**Assumes feed grain decline falls on animal producers

Table 5. Effects of drought-induced crop production declines on total production in MINK.

estimating intersectoral multiplier effects of drought-induced changes in agricultural production are discussed. In the MINK study, only corn and sorghum multipliers were considered because of their strong linkage to animal production, which in turn is strongly linked with the rest of the regional economy through the large meat-packing industry.

Two sets of estimates of the multiplier effects of drought-induced production declines in corn and sorghum are shown in Table 5. In one set, it is assumed that rising production costs of feed will result in a reduction in export demand for corn and sorghum that is proportional to the decline in production (Bowes and Crosson, 1991). Animal producers in the MINK region would presumably import lower-priced grain from outside the region to minimize their own production costs. In the other set of estimates, it was assumed that exports would remain at predrought levels and that the region's animal producers would bear the brunt of grain production declines in the form of higher feed costs. Easterling et al. (1992) point out that neither of these two sets are realistic since higher MINK feed production cost would result in lower exports and higher animal production costs simultaneously.

The worst-case short-term drought scenario—no adjustments and feed grain declines are borne by animal producers—results in a 9.7% decline in total regional production. The best-case long-term drought scenario—with adjustments and grain production declines fall on exports—results in only a 0.5% decline in total regional production. Bowes and Crosson (1991) argue that the reason for the differences in these two extreme cases is that when the decline in grain production is absorbed by animal producers, animal production declines. Since animal production is strongly linked to the meat-packing industry, which in turn is a strong multiplier to the regional economy, declines in animal outputs will sharply reduce overall regional product.

The low-end estimates of the multiplier effects of declining grain production are most likely for two reasons. First, farmers can be expected to make adjustments to extended drought. Second, it does not make sound economic sense for grain production shortfalls to be absorbed totally by animal producers in order to maintain exports. Not only could MINK animal producers import relatively less expensive grain from outside the region, but MINK grain exports would be relatively more expensive on the open markets because of the drought-induced rise in production costs.

SUMMARY AND CONCLUSIONS

The estimation of the regional consequences of drought requires the integration of a hierarchy of models to describe the primary impacts at the farm/firm level, the secondary impacts at the regional level, and the linkages between affected economic sectors. The models that belong in such a hierarchy have been developed but not integrated to perform regional analyses. In this chapter, a methodology that was developed to examine the regional impacts of and responses to long-term greenhouse warming-induced climate change is applied to the estimation of regional consequences of widespread and persistent drought.

The methodology was developed in a case study of the Missouri-Iowa-Nebraska-Kansas (MINK) region. Although all important resource sectors were studied in the project, only water resources and agriculture were discussed here as an illustration of the methodology. To demonstrate the methodology, a hypothetical recurrence of the drought conditions of the 1930s in MINK was examined. To emulate the impacts of short-term (one to three years) drought, adjustments were not made to production operations in response to drought conditions because of inadequate lead times. In the case of long-term drought (four years or longer), it was assumed that adjustments would be made to adapt production operations to the drought conditions.

The study proceeded with a baseline description of the MINK economy, showing the specialization of MINK in agriculture. A farm-level analysis was performed with the EPIC crop simulation model to determine the impact of the 1930s drought conditions on current crop productivity. Under both short- and long-term drought conditions, dryland corn and sorghum yields were substantially reduced. Dryland wheat was essentially unaffected by the drought conditions. Irrigation on demand gave strong protection to all crops against the drought conditions. However, it is not likely that the increased irrigation requirements induced by the drought conditions could be met on a sustained basis. Thus, declines in irrigated acreage were assumed.

Farm-level simulations of crop yield response to the 1930s drought conditions were scaled up to the regional level and used to calculate regional changes in agricultural production. Drought-induced regional production declines translated into percentage declines in the upper teens in the total regional value of agricultural output in MINK.

Finally, the IMPLAN input-output model was used to quantify the multiplier effects of declining feed grain production on total regional production in MINK. Were farmers able to freely import relatively cheaper grain from nondrought areas while at the same time reducing exports, the impact of the drought conditions on the overall MINK economy probably would not be large. Were farmers to try to maintain current export levels, the relatively high-price feed grain that is the residual of exports would cause as much as a 10% decline in total regional production in MINK. In both cases, the agricultural component of the MINK economy would suffer disproportionate declines from drought.

Future Directions

The MINK methodology can serve as a model for the integrated assessment of the regional consequences of drought. The methodology can, however, be improved. In the case of short-term drought assessment, the EPIC model should be given enough predrought growing seasons to stabilize yield estimates before encountering a drought year. This will eliminate the need to use a long-term drought as a surrogate for a short-term drought, as was reported above.

Explicit attention was not given to climatic and economic conditions in neighboring regions, which either complement or compete with the MINK region. Such conditions should be given attention in future analyses.

Finally, the role of government intervention in both short-term and long-term droughts is not taken into account. Government transfer payments can make a large difference in the degree of economic hardship a region must endure during droughts. Future analyses should acknowledge the potential of the government to provide drought relief.

Despite these deficiencies, the MINK methodology provides a foundation to build on the above-mentioned improvements. The methodology is transportable to regions in which there are sufficient input data sets for the various embedded models and estimation techniques.

ACKNOWLEDGMENTS

This research is part of an effort sponsored by the U.S. Department of Energy, Office of Health and Environmental Research. Several people contributed to the work who do not appear as authors of this chapter: Norman J. Rosenberg, Pierre R. Crosson, Michael D. Bowes, Mary S. McKenney, and Kenneth D. Frederick.

ABOUT THE AUTHOR

William E. Easterling is an assistant professor of agricultural meteorology at the University of Nebraska-Lincoln and the director of the Great Plains Regional Center for Global Environmental Change. His research interest focuses on the interaction of climate variability and change with natural resources, especially agriculture and water. Specifically, he studies the sustainability of alternative patterns of natural resource management under climate change in marginal ecosystems. He is also interested in the role of climate in economic decision making.

REFERENCES

Bowes, M. D.; and P. Crosson. 1991. Consequences of climate change for the MINK economy: Impacts and responses. Report VI in *Processes for Identifying Regional Influences of and Responses to Increasing Atmospheric CO_2 and Climate Change—The MINK Project*. U. S. Department of Energy Report TR052H, DOE/RL/01830-H12.

Easterling, W. E.; P. Crosson; N. J. Rosenberg; M. S. McKenney; and K. D. Frederick. 1992. Methodology for assessing regional economic impacts of and responses to climate change: The MINK study. *In* J. Reilly and M. Anderson, eds. *Economic Issues in Global Climate Change: Agriculture, Forestry and Natural Resources*; pp. 168-99. Westview Press, Boulder, Colorado, U.S.A.

Frederick, K. D. 1991. Water resources. Report IV in *Processes for Identifying Regional Influences of and Responses to Increasing Atmospheric CO_2 and Climate Change—The MINK Project*. U. S. Department of Energy Report TR052F, DOE/RL/01830-H10.

Harbeck, G. E., Jr. 1962. *A Practical Field Technique for Measuring Reservoir Evaporation Utilizing Mass-Transfer Theory*. Geological Survey Professional Paper 272-E, U.S. Government Printing Office, Washington, D.C.

Kromm, D. E.; and S. E. White. 1986. Variability in adjustment preferences to groundwater depletion in the American High Plains. *Water Resources Bulletin* 22(5):791-801.

Nellis, M. D. 1987. Land-use adjustments to aquifer depletion in western Kansas. *In* C. Cocklin, B. Smit, and T. Johnston, eds. *Demands on Rural Land: Planning for Resource Use*; pp. 71-84. Westview Press, Boulder, Colorado, U.S.A.

Penman, H. L. 1948. Natural evaporation from open water, bare soil and grass. *Proceeding of the Royal Society of London, Series A* 193:120-45.

Riebsame, W. E.; S. A. Changnon, Jr.; and T. R. Karl. 1990. *Drought and Natural Resources Management in the United States: Impacts and Implications of the 1987-89 Drought*. Westview Press, Boulder, Colorado, U.S.A.

Rosenberg, N. J., ed. 1980. *Drought in the Great Plains: Research on Impacts and Strategies*. Water Resources Publications, Littleton, Colorado, U.S.A.

Rosenberg, N. J.; and P. Crosson. 1991. The MINK Project: A new methodology for identifying regional influences of and responses to increasing atmospheric CO_2 and climate change. *Environmental Conservation* 18(4):313-22.

U.S. Forest Service. 1989. *Micro IMPLAN Release 89-03 Help File*. U.S. Department of Agriculture, Forest Service, Land Management Planning, Fort Collins, Colorado, U.S.A.

Wilhite, D. A.; and W. E. Easterling, eds. 1987. *Planning for Drought: Toward a Reduction of Societal Vulnerability*. Westview Press, Boulder, Colorado, U.S.A.

Williams, J. R.; C. A. Jones; and P. T. Dyke. 1984. A modeling approach to determining the relationship between erosion and soil productivity. *Transactions of the American Society of Agricultural Engineers* 27:129-44.

Yevjevich, V.; W. A. Hall; and J. D. Salas. 1978. *Drought Research Needs*. Water Resources Publications, Fort Collins, Colorado, U.S.A.

Chapter 5
Agricultural Drought Management Strategies to Alleviate Impacts: Examples from the Arid and Subhumid Regions of the Indian Subcontinent

A. S. R. A. S. Sastri

INTRODUCTION

Drought, though not considered a disaster by many, is one of the natural calamities encountered by humans. Drought may begin at any time, last indefinitely, and attain many degrees of severity. Its beginning is subtle, its progress is insidious, and its effects can be devastating (World Meteorological Organization [WMO], 1975). Often, drought is associated with aridity, although in reality it is a climatic aberration that can occur in any part of the globe, including humid and per humid regions. It is a condition relative to some long-term average of balance between rainfall and evapotranspiration in a particular area (Wilhite and Glantz, 1985).

Droughts are not quick-onset disasters, like floods or typhoons, but are usually the eventual result of a set of weather sequences that require extensive periods of time to develop (Linsley et al., 1959). Thus, according to Wilhite (1982), drought is a "creeping phenomenon," the effects of which accumulate slowly over a considerable period of time. However, drought means different things to different people and no universally accepted definition has so far been developed (Sastri et al., 1982).

Droughts are classified in three main categories (Das, 1981): meteorological (rainfall substantially below climatological expectations); hydrological (surface and subsurface water [i.e., reservoirs, lakes, rivers, and groundwater] depleted); and agricultural (inadequate soil moisture to support the growth of plants and crops to maturity).

AGRICULTURAL DROUGHTS—GENERAL CONCEPTS

Agricultural drought is generally accepted to be a period of dryness caused by a precipitation shortfall during the crop growth period (Huda et al., 1986), resulting in a substantial reduction in yield. The impacts of drought depend not only on the duration and intensity but also on the crop, its stage of growth, and soil characteristics (e.g., water-holding capacity). Thus, agricultural drought is a crop-specific phenomenon.

In the *Technical Report on Drought and Agriculture*, the Commission for Agricultural Meteorology (CAgM), a working group of the World Meteorological Organization (WMO, 1975), gave a detailed account of the concept of agricultural drought and several meteorological indices used for detecting severity and impact. In the discussion that follows, the concept of agricultural drought is discussed in reference to the arid and subhumid regions of the Indian subcontinent.

CLASSIFICATION OF AGRICULTURAL DROUGHTS

Knowledge of the frequency and time of occurrence of agricultural droughts of varying intensity is of paramount importance in identifying crops best suited to a given region (Sastri et al., 1981). Some attempts have been made to classify agricultural drought using water availability periods and monthly rainfall deciles (George and Krishna, 1969; George and Kalyana Sundaram, 1969) for some parts of India and using an aridity index for the Indian arid region (Krishnan and Thanvi, 1971). However, these are general in nature and not crop-specific classifications.

Sastri et al. (1981b) developed a new method for classifying agricultural droughts based on a rational approach for pearl millet and wet season (kharif) pulses for the arid region of western India. The index of moisture adequacy (IMA), a ratio of actual to potential evapotranspiration during the crop growth period, is the basis for this classification. However, the drought or water stress period at different phenological phases of crop growth affects crop productivity differently. Ramana Rao et al. (1981) developed an integrated scheme of drought classification in which the values of IMA during the seedling, vegetative, and reproductive stages were considered. Water stress at these three stages was integrated with different combinations of drought intensity at different growth stages in order to assess the overall intensity of drought for the arid regions of India. More recently, Sastri and Patel (1984) developed another classification of agricultural droughts for rainfed rice, grown extensively in the subhumid tracts of central India.

Indices for Agricultural Drought Classification

Agricultural droughts have been classified according to degrees of water availability such as humid, moist, and submoist (Cocheme and Franquin, 1967). For the Indian arid region, Krishnan and Thanvi (1971) used the aridity index (IA) of Thornthwaite (1948). For studies of drought occurrence, this index is a ratio (%) of annual water deficit to the annual potential evapotranspiration.

For their crop-specific classification, Sastri et al. (1981b) and Ramana Rao et al. (1981) used the IMA, developed by Subrahmanyam and Subramaniam (1964). This index has also been referred to as the R-Index (Yao, 1969). The IMA is widely used for classifying agricultural droughts in many parts of the world.

Evaluation of the Minimum Required Value of the Index

Weekly values of IMA have been computed with the help of a climatic water balance using the bookkeeping procedure of Thornthwaite and Mather (1955). To calculate the water balance, potential evapotranspiration values were computed using the Penman (1948) equation. The average values of IMA during the crop season have been determined for a number of years for which crop productivity data are available. The minimum required value of IMA for obtaining the average productivity for a particular crop for a particular area/region was calculated with the help of graphical interpretation (Figure 1). This is similar to the method suggested by Azzi (1956). Using this method, the average value of IMA for that crop for each year is marked with either an open circle (for below-average productivity) or a closed circle (for above-average productivity), depending on its productivity in that particular year. A line is drawn parallel to the x-axis, keeping the closed circles above the line and the open circles below the line. If deviations occur, care is taken to see that an equal number of deviations occur on either side of the line. Thus, the line represents the minimum required value of IMA for obtaining the average productivity of the crop in that area.

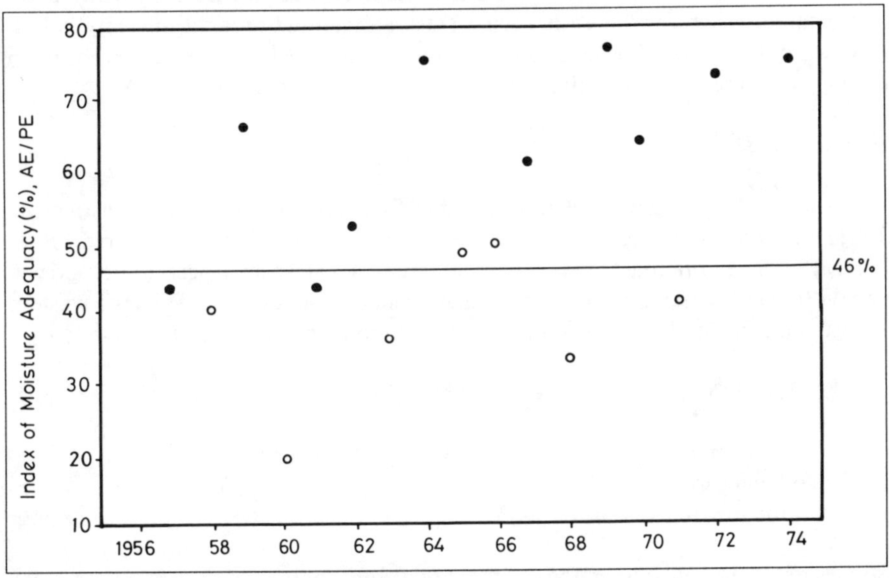

Figure 1. Minimum required value of IMA for obtaining above-average yield for pearl millet at the Jodhpur station.

AGRICULTURAL DROUGHT CLIMATOLOGY

In India, long-term time series of crop productivity data are available for different crops in different districts of each state. Similarly, time series of rainfall and other meteorological data are also available. Using time series of crop productivity and rainfall and other meteorological data, the agricultural drought climatologies of some crops for different areas or districts of the arid and subhumid regions of India were studied. Such studies are very useful in identifying crops and areas within a region that are susceptible to drought. These studies also help to assess the impact of drought of different degrees of intensity on crop productivity. Such specific case studies for arid and subhumid climatic regions of India are discussed below.

INTEGRATED SCHEME OF AGRICULTURAL DROUGHT CLASSIFICATION

It is well known that even with identical moisture deficits, the intensity of drought for a given crop will depend on the phenological stage of the crop when the stress occurs. Moreover, stress conditions of varying intensities may occur at different growth stages in the same season for the same crop. Thus, agricultural drought intensity is best assessed if the degree of water stress and its duration is analyzed for each of the main crop growth stages (i.e., seedling, vegetative, reproductive, and maturity).

Arid Regions

Agricultural droughts have been classified for pearl millet and pulses (Sastri et al., 1981), the two major crops in the Indian arid regions. The minimum required values of AE/PE for determining the average productivity with respect to pearl millet and wet season (kharif) pulses for three representative stations in the arid region of western Rajasthan, obtained from the graphical interpolation method, are as follows:

STATION	Pearl Millet	Kharif Pulses
Sikar (semiarid)	6	68
Jodhpur (arid)	46	55
Barmer (extremely arid)	39	55

Studies of the drought climatology of these two crops for three representative stations revealed that kharif pulses are more susceptible to drought than pearl millet. Since the maturation period for kharif pulses is short, water stress during the crop growing season is often avoided.

IMA during Different Crop Growth Stages (%)	Drought	Seedling Stage	Vegetative Stage	Reproductive Stage
76 - 100	None	S0	V0	R0
51 - 75	Mild	S1	V1	R1
26 - 50	Moderate	S2	V2	R2
25 or less	Severe	S3	V3	R3

Table 1. The range of IMA values for different intensities of drought during the crop growth stages of pearl millet.

The above studies are based on an average value of IMA during the entire crop growth season. An integrated drought classification scheme was developed by Ramana Rao et al. (1981) that considered values of IMA during each crop growth stage. This scheme was slightly modified by Sastri et al. (1984) and used to study the productivity pattern of different crops. Codes of S, V, and R were given for each of the growth stages (seedling, vegetative, and reproductive, respectively), and the intensity during each stage was suffixed as 0, 1, 2, and 3. A specific example for pearl millet for the arid region is given in Table 1.

For pearl millet with a 14-week maturation period, agricultural drought with different intensities is worked out using combinations of intensity at different growth stages (see Table 2), as follows:

1. Severe drought for a pearl millet crop (A3) when both vegetative and reproductive stages experience severe drought (V3 and R3 conditions) with any combination of S, such as S0, S1, S2, or S3.

2. Moderate drought for pearl millet (A2) when vegetative and reproductive stages experience one moderate and one severe drought (combinations such as V2R3 or V3R2) with any combination of S.

Station	Percentage Probability of:			
	No Drought (A0)	Mild Drought (A1)	Moderate Drought (A2)	Severe Drought (A3)
Sikar	31	44	3	22
Jodhpur	16	28	16	40
Nagaur	7	43	11	39
Barmer	7	46	7	40
Jaisalmer	1	11	9	79

Table 2. Percentage probabilties of various intensities of drought for pearl millet in different stations of Western Rajasthan.

3. No drought conditions for pearl millet crop (A0) when mild drought prevails at one or two stages of crop growth with no drought in the third stage (conditions such as S0V1R1, S0V0R1, S0V1R0, S1V0R0, S1V1R0, S1V0R1).

4. Mild drought conditions for the other combinations.

Based on this methodology, the probability of various intensities of drought for the pearl millet crop was calculated for different stations in the arid region of western Rajasthan (Sastri et al., 1982). The results of this analysis are shown in Table 2.

If mild and no-drought conditions are considered favorable for crop production and moderate and severe drought conditions are considered unfavorable, then the probability of favorable conditions for pearl millet growth is higher at Sikar (75%) than at Barmer (53%). In the Jodhpur region, the probability for unfavorable conditions is greater (56%), and it is very high at Jaisalmer (88%). In Nagaur, it is even, with a 50% probability for favorable and unfavorable conditions for pearl millet growth.

The productivities of pearl millet and kharif pulses under no-drought, mild, moderate, and severe drought conditions are shown in Figure 2. Except under moderate drought conditions, the productivity of pearl millets is always higher than that of kharif pulses. But the probability of moderate drought for pearl millets is comparatively lower. Thus, the productivity of pearl millet under a given intensity of drought is always higher than that of kharif pulses. However, when crop economics are considered, kharif pulses are more economical than pearl millet because their market price is nearly double that of pearl millet.

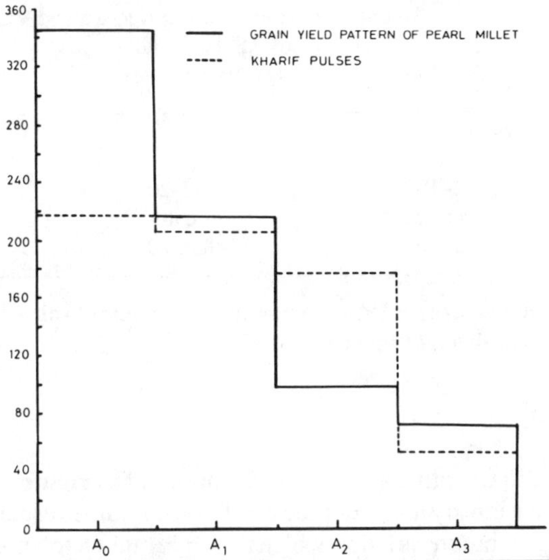

Figure 2. Grain yield pattern of pearl millet (solid line) and kharif pulses (dashed line) during different intensities in western Rajasthan.

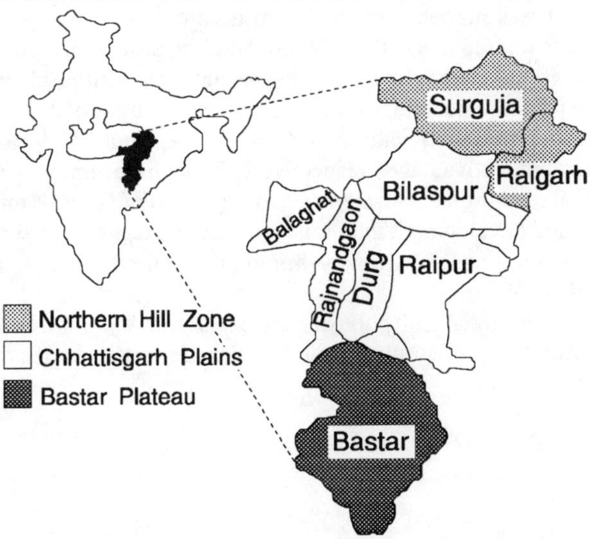

Figure 3. The Chhattisgarh region and its agroclimatic zones.

Intensity	Departure from Minimum Required Value of AE/PE during: Seedling	Vegetative	Reproductive
Mild	1-5	1-10	1-10
Moderate	6-10	11-20	11-20
Severe	Greater than 10	Greater than 20	Greater than 20

Table 3. Range of departures of IMA from minimum required values at different growth stages under different drought intensities.

Subhumid Region

In the subhumid Chhattisgarh region of central India (Figure 3), rice is grown on about 4.2 million ha, mostly under rainfed conditions. Farmers usually grow photosensitive tall and long-duration varieties, which flower in mid-October and mature by mid-November. The southwest monsoon withdraws by mid-September. Thus the rice crop is usually grown without rains for about two months, which often results in drought conditions during the reproductive stage. The success of the rice crop depends mostly on the cyclonic activity in the Bay of Bengal in the month of October. Intermittent drought due to a break in monsoon conditions is common.

Working with rainfed rice in the subhumid region, Sastri and Patel (1984) worked out a similar integrated scheme for classifying drought. Under this scheme, water stress during different physiological stages was classified on the basis of the departure of IMA from the minimum required values of 75%, 100%, and 100% during seedling, vegetative, and reproductive stages, respectively. Though the minimum required values of IMA are less at seedling stage, water stress during seedling stage is more detrimental to crop growth and development than it is in the other stages. Based on this criteria, the range of departures of IMA from the minimum required values at different stages were calculated (Table 3).

Thus, for a traditional tall photosensitive variety, *Safri*, with about 140 days required for maturation, agricultural drought intensities are classified during the three growth stages as follows:

Crop Stage	Duration (weeks)	Minimum Value of AE/PE	AE/PE Range for Drought Intensity of:		
			Mild	Moderate	Severe
Seedling	3	75	70-74	65-69	< 65
Vegetative	7	100	90-99	80-89	< 80
Maturity	5	100	90-99	80-89	< 80

Based on this classification, the number of weeks under each category of drought during the three physiological stages was computed. To integrate drought for the entire crop season, mild, moderate, and severe droughts are ranked 1, 2, and 3, respectively. To obtain the total rank for the entire season, the drought intensity code at each crop stage was multiplied by the number of weeks at that stage and summed.

The integrated drought code (IDC) was computed from:

$$IDC = \sum_{i=1}^{n} C_i W_i$$

C_i = Drought code for the physiological stage i
W_i = Number of weeks under C_i

Using these IDC values, the drought intensity was classified as:

Drought Intensity Code		Integrated Number
None	0	0 - 3
Mild	1	4 - 6
Moderate	2	7 - 9
Severe	3	> 9

No upward trend has been apparent in the productivity of rice in this region for several decades (Table 4). However, considerable technological changes have occurred, such as the development of high-yielding varieties (HYVs), irrigation, and fertilizers. Based on the classification of agricultural drought for rice, the years of severe drought and normal rainfall were compared (Figure 4). Figure 4 illustrates that at a given moisture status, productivity in recent years is increasing. This can be attributed to technological developments.

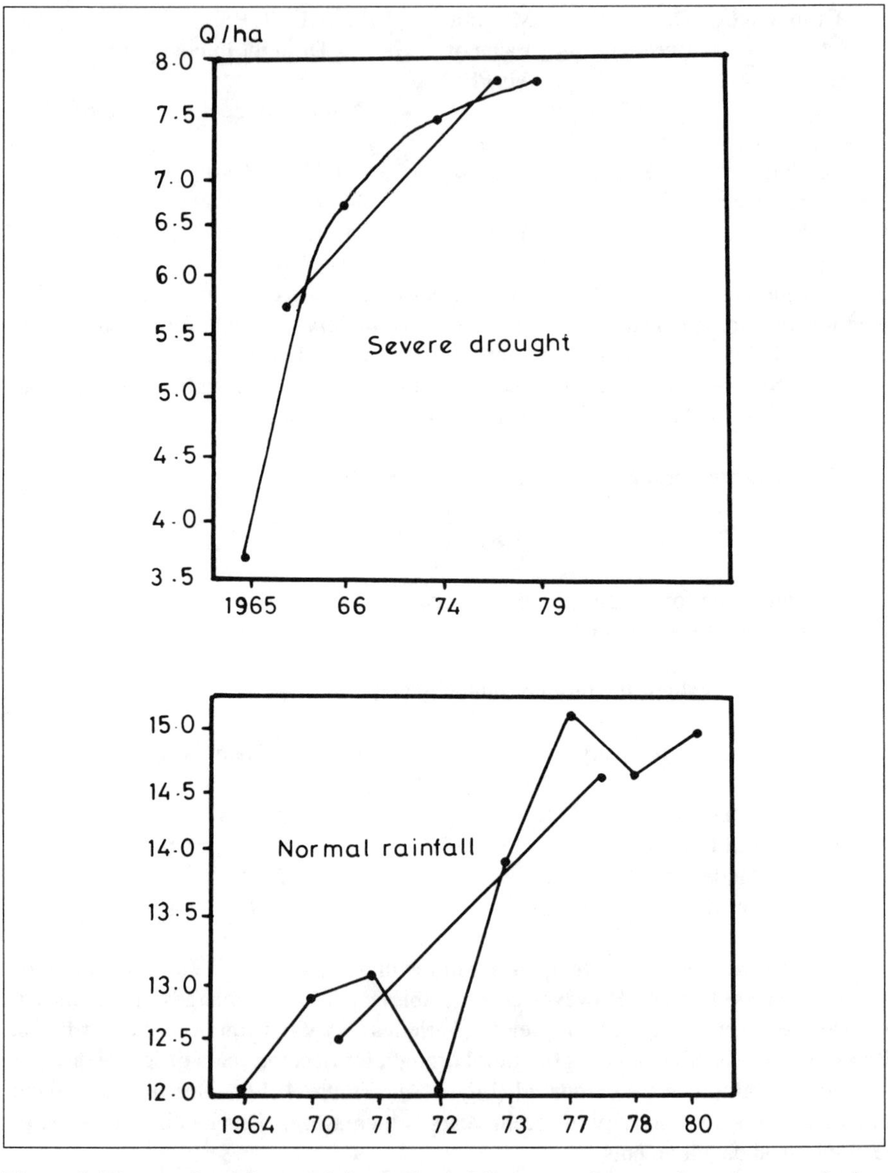

Figure 4. Rice productivity trends in good rainfall years and severe drought years during the period 1964-81 in Raipur.

Station	Average Productivity of Rice (t/ha) during: 1950-60	1961-70	1971-80
Dhamtari	—	1.04	0.85
Balodabazar	—	0.74	0.89
Bhatapara	—	0.71	0.79
Mahasamund	0.94	0.95	1.04
Pithora	1.04	0.93	0.93
Gariabund	0.82	1.18	0.99
Deobhog	0.88	0.96	0.91
Average	0.92	0.93	0.91

Table 4. Decadal average productivity of rice at a few stations in Raipur district.

STRATEGIES FOR ALLEVIATING DROUGHT IMPACT

To alleviate the impact of drought on agricultural crops, several strategies, such as moisture conservation or altering the cropping pattern, can be adopted. In this section, a few case studies of arid and subhumid regions of India are highlighted.

Water Harvesting

One important technique for alleviating the impact of drought on agricultural crops is the adoption of water-harvesting techniques through which excess rainwater (runoff) is collected, conserved, and recycled at the time of stress for agricultural crops.

For the Indian arid regions, the water-harvesting potential for two districts (Jodhpur and Nagaur) was calculated using the climatic water balance approach (Sastri et al., 1982). It was determined that the eastern parts of these districts have the greatest potential for water harvesting. In these regions, the average surplus water varied from 50 to 100 mm in Jodhpur district and above 100 mm in Nagaur district in August and August and September, respectively.

For the subhumid regions of central India, rainwater management is more important because much water is lost either through runoff or through deep percolation during the active monsoon months. Sastri and Urkurkar (1989) have calculated the collection required for runoff water to meet the deficiency for rice in order to avoid water stress. Total daily water loss through evapotranspiration and percolation was assumed to be about 10-12 mm. Thus, the total water requirement for a medium-duration rice

District	Rain Water Needed to be Harvested for Rice Area (cu. m)	Dimensions (L x B) of Farm Pond with 3-meter Depth:	
		Length (m)	Breadth (m)
Raipur	2,230	27	27
Durg	4,796	40	40
Rajnandgaon	4,782	40	40
Bilaspur	2,804	31	31
Balaghat	2,223	27	27
Kanker Tehsil	3,452	33	33
Raigarh and Sarangarh Tehsils	1,314	21	21

Table 5. Area required for water harvesting in different districts in Chhattisgarh Plains agroclimatic region of central India.

crop is 1,500 mm. The corresponding dimensions of farm ponds were also calculated (Table 5).

Sastri and Urkurkar (1989) also worked on an in situ water-harvesting technique with a series of raised and sunken beds (rainwater cistern system). Rice is grown in the sunken bed and soybean in the raised bed. Excess rainwater draining from the raised bed to the sunken bed thus provides extra water to the rice crop. The results of a three-year experiment are given in Table 6. It was concluded that because of the large amount of water necessary for rice, the size of the farm pond necessary to collect runoff water was prohibitive. The authors concluded that the farm pond system is not economical in these cases.

More recently, Sastri and Chaudhary (1991) concluded, using the water balance for a bunded[1] rice field under three major soils (sandy loam, clay loam, and clayey) during excess, normal, and deficit rainfall patterns, that in the bunded rice fields, percolation losses are very high compared to runoff losses in these regions. It was suggested that "dug" or "dug cum bore wells" may be more useful than farm ponds in providing life-saving irrigation to rainfed rice crops. The details of the water balance are given in Table 7.

[1]Rice fields are usually bunded according to the slope to make the water level uniform in the fields. In the Chhattisgarh region of central India, rice fields are usually bunded (diked) 1.5-2.0 m high to impound water up to 30-45 cm because of the risk of drought at later stages.

Year	Seasonal Rainfall (mm)	Ineffective Rainfall (mm)	Surplus Water (mm)	Total Runoff (mm)	Water Available: Soybean (mm)	Water Available: Rice (mm)	Ratio of Runoff to Rainfall (%)	Grain Yield (t/Ha): Soybean RCS	Grain Yield (t/Ha): Soybean Plain	Grain Yield (t/Ha): Rice RCS	Grain Yield (t/Ha): Rice Plain
1981	1,119	317	196	513	607	1,632	46	2.2	1.4	2.8	2.0
1982	792	210	166	376	416	1,168	47	2.2	1.3	2.2	1.2
1983	1,263	335	234	568	695	1,831	45	2.4	1.7	3.0	1.4

Table 6. Water harvesting and productivity pattern in the rainwater cistern system (RCS) during different years (Sastri and Urkurkar, 1989).

District	Parameter (mm)	Sandy Loam Soils Excess*	Normal*	Deficit*	Clay Loam Soils Excess	Normal	Deficit	Clayey Soils Excess	Normal	Deficit
Raipur	Rainfall	1,813.5	1,142.2	763.2	1,813.5	1,142.2	763.2	1,813.2	1,142.2	763.2
	Percolation	1,032.6	434.9	137.1	730.8	389.6	76.1	220.2	312.6	26.3
	Runoff	85.1	0.0	0.0	262.9	0.0	0.0	398.3	27.2	0.0
Rajnandgaon	Rainfall	2,002.3	1,232.2	682.4	2,002.3	1,232.2	682.4	2,002.3	1,232.2	682.4
	Percolation	958.0	535.8	59.2	850.7	475.8	8.3	621.6	405.7	0.0
	Runoff	75.2	0.0	0.0	323.1	10.0	0.0	724.1	29.7	0.0
Durg	Rainfall	1,736.5	1,136.7	862.7	1,736.7	1,136.7	862.7	1,736.7	1,136.7	862.7
	Percolation	981.2	450.5	244.8	696.5	390.5	193.8	580.5	314.8	137.6
	Runoff	0.0	0.0	0.0	0.0	0.0	0.0	375.7	24.8	6.2
Bilaspur	Rainfall	1,475.6	1,201.0	804.3	1,475.6	1,201.0	804.3	1,475.6	1,201.0	804.3
	Percolation	970.9	462.1	153.2	499.4	390.1	103.2	399.4	348.3	52.7
	Runoff	18.5	0.0	0.0	143.8	22.0	0.0	244.4	13.9	0.0

* Excess = Greater than M + S.D. Normal = M Deficit = Less than M - S.D.
Where M = Mean, S.D. = Standard Deviation

Table 7. Water balance of bunded rice field in different soil types and in different rainfall patterns at four districts in Chhattisgarh plains region (Sastri and Chaudhary, 1991).

Onset of Monsoon	Period	Average Rainfall (mm)	Water Availability Periods: Moist I (PE>R>PE/2)	Humid (R>PE)	Moist II (PE>R>PE/2)	Rice Productivity (t/Ha)
Early	Before 6/11	1,607	7	127	12	1.18
Normal	6/11-6/30	1,366	14	115	10	1.15
Late	Later than 6/30	1,044	13	91	14	0.84

Table 8. Effect of onset of monsoon on the length of growing season and rice productivity in Raipur District (database 1941-88).

Adjusting Crops and Cropping Patterns

It is well known that in any given year, the impacts of drought differ from crop to crop. Based on some experimental results, the effect of moisture distribution during the growing season on the productivity of different crops was worked out by Patel et al. (1986). In this study, the impact of drought on rice, soybeans, blackgram, groundnut, and minor millets was examined at four different crop growth stages (seedling, vegetative, reproductive, and maturity) under the same moisture distribution (expressed as AE/PE [%] in the years 1981 and 1982), as shown in Figure 5. It was determined that water stress is more detrimental at seedling stage than at vegetative stage. From a water stress point of view, rice at Raipur is more drought-prone than groundnuts and soybeans. Based on the effect of water stress at different growth stages on the productivity of these crops, the most economically viable crops can be chosen for different seasonal moisture distribution patterns.

Manipulation of Crop Varieties

The success of a crop and crop variety, especially under rainfed conditions, depends on the onset and duration of the monsoon. The effect of monsoon onset on the duration of water availability periods and corresponding rice crop productivity in the Raipur district of central India was analyzed by Sastri and Verma (1985) and is shown in Table 8. Late onset of monsoon results in late sowing of the crop and hence in drought conditions during reproductive and maturity stages. Therefore, if the monsoon onset is delayed, it is necessary to manipulate crop varieties or the irrigation source.

Using a systems approach, Sastri et al. (1990) conducted an experiment in which different approaches for drought management were evaluated, including varietal adjustment. It was found that during the drought year of 1988 at Raipur, medium duration rice yielded 1.56 t/ha, compared to a long-duration variety yield of only 0.82

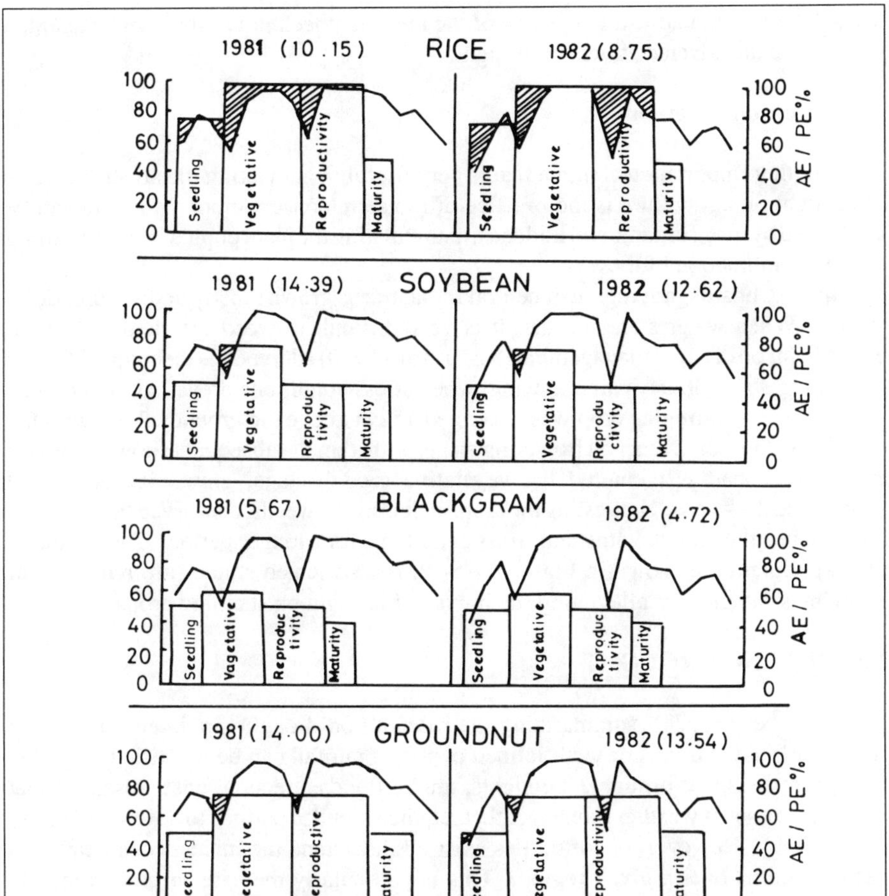

Figure 5. Drought pattern during 1981 and 1982 in Raipur region.

t/ha. Thus varietal adjustment is one of the important techniques that can be adapted to alleviate the adverse effects of drought.

Moisture Conservation

Another important technique that is very useful in alleviating the adverse effects of drought on agriculture is the practicc of moisture conservation. Moisture can be conserved by using organic soil mulches, manipulating the plant population, or weeding (or any combination of these).

In the Chhattisgarh region of central India, rice is grown mostly under a broadcast system. When water stress occurs, it is very difficult to weed the fields. In their experiment on drought management, Sastri et al. (1990) adopted the technique of weed harvesting about 30 days after sowing, when stress conditions prevailed. Under this technique, the entire rice crop was harvested 15 cm above the ground, along with the weeds. As the weeds reached flowering stage, with a small subsequent shower, the rice crop, which had just entered the vegetative growth stage, grew vigorously and suppressed the weeds. Harvesting the crop, including the weeds, conserved the moisture and allowed the crop to withstand stress conditions for a longer period. The results of the experiment are shown in Figure 6. Such field-oriented studies are required for evolving strategies for alleviating the impact of drought on standing crops.

PREDICTING DROUGHT

Landsberg (1974), summarizing the research on droughts, pointed out that no radical one-sided trends or well-defined cycles of rainfall can be identified to enable reliable forecasts of drought. Droughts, unlike floods, are slow-onset disasters that result from a set of weather sequences that require extended periods to develop (Linsley et al., 1959). However, opportunities exist to understand the incidence, spread, and decay patterns over a given region. This is particularly the case in predominantly monsoonal climates where rainfall is a result of a weather disturbance usually originating in the ocean. Subrahmanyam and Sastri (1971) have been successful in predicting the origin of drought in the Deccan plateau. Ramakrishna and Sastri (1980), Sastri and Malakar (1981), Sastri and Ramakrishna (1981), Sastri et al. (1981c), and Ramakrishna et al. (1984) have also attempted to predict drought over the arid region of western Rajasthan. Sastri (1988) undertook similar research for central India.

The studies for western Rajasthan revealed that drought has a general tendency to originate in the northeastern region around Churu and Jhunjhnu during July, spread in a southwesterly direction into Barmer during August, and dissipate with an easterly movement into the Pali region in September. Contrary to this, in central India (Madhya Pradesh state) it was observed that drought conditions are discernable initially at

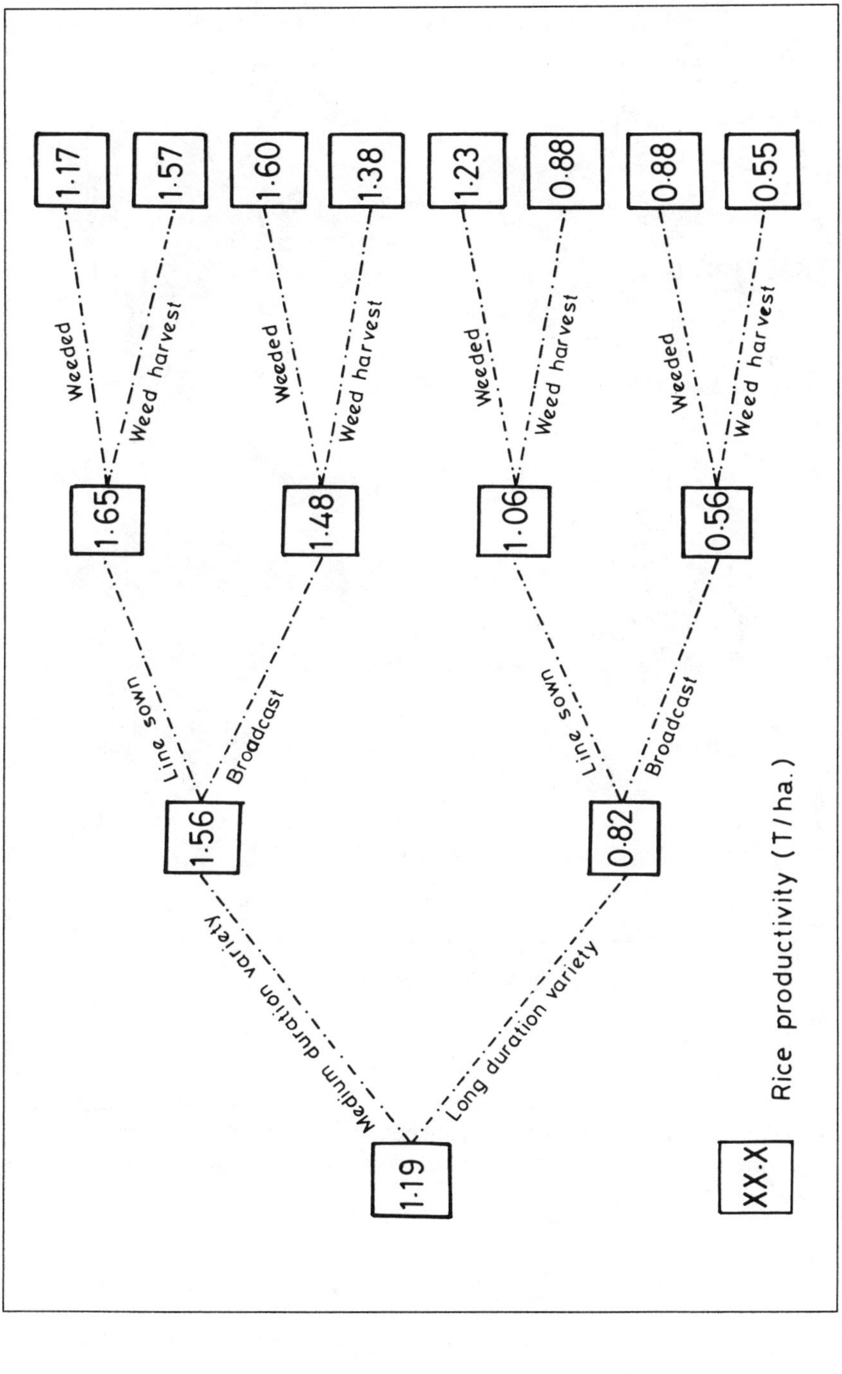

Figure 6. Systems analysis for drought management of rice in central India.

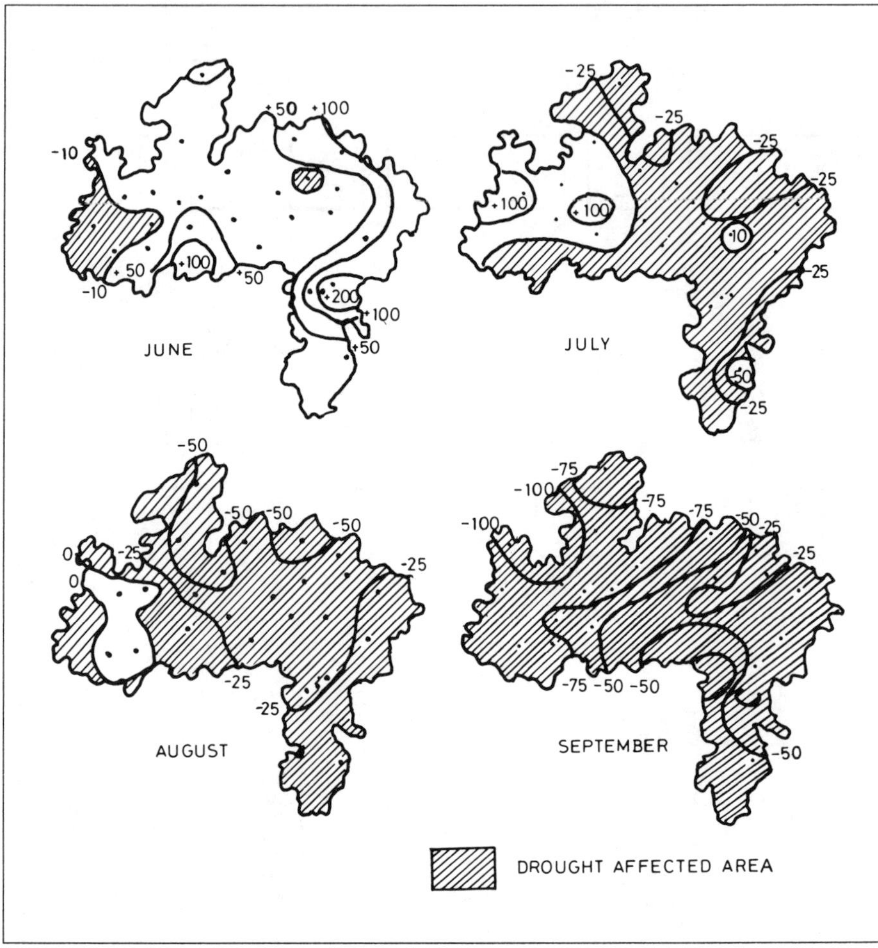

Figure 7. Drought during 1986 over Madhya Pradesh.

Jagadalpur, the starting point of the southwest monsoon for the state; they then spread toward the north and diminish with a westerly movement in the subsequent months (Figure 7). Such studies on synoptic drought climatology are very important because they help in preparing contingency plans to alleviate drought impacts, especially in the agricultural sector.

SUMMARY AND CONCLUSIONS

Although several classifications of agricultural drought exist, they are mostly general in nature and are not crop specific. Hence, crop specific agricultural droughts were classified on the basis of the minimum required values of the index on moisture adequacy (IMA) or AE/PE.

Because the same quantity of moisture deficiency may result in different degrees of drought intensity for the same crop at different stages, an integrated scheme of drought classification was developed based on the values of IMA at seedling, vegetative, and reproductive stages of crop growth for pearl millet and wet season (kharif) pulse crops in the arid region and for rainfed rice crops in the subhumid region of India.

Based on the integrated scheme of drought classification, the probable favorable conditions for pearl millet and kharif pulse crops were examined. It was found that the productivity of pearl millet under a given intensity of drought is always higher than that of kharif pulses. However, the market prices of these two crops are different, and therefore kharif pulses may be more economical than pearl millet.

In the case of rainfed rice crops in subhumid regions of India, based on the integrated scheme of drought classification, it was found that at a given degree of drought, the productivity of rice in recent years is increasing. This can be attributed to technological changes such as HYVs, irrigation, chemical fertilizers, and so on.

To alleviate the impact of drought, experiments were conducted on a rainwater cistern system with a series of sunken and raised beds, with rice grown in sunken beds and soybeans grown in raised beds. The success of the experiment led to the conclusion that wherever adequate drainage facilities are available under field conditions, soybean crops can be taken and the drained water from the soybean field can be used in the rice fields. Also based on climatic water balance of bunded rice fields, it was found that the percolation losses are very high compared to runoff losses, and it was suggested that "dug" or "dug-cum-bore wells" may be more useful than farm ponds in providing life-saving irrigation to rainfed rice crops.

The crops and cropping pattern can also be adjusted to alleviate drought impacts with a knowledge of the performance of different crops in stress conditions at different stages of growth. Similarly, varietal adjustment with proper moisture conservation techniques can also be done to alleviate the impact of drought.

It is clear that droughts do not descend suddenly, like floods do, but are slow-onset disasters that follow a set of weather sequences. Based on the climatological studies on drought prognostication, especially in a monsoon climate, the origin, spread, and decaying pattern of drought in a given region can be understood, which helps in preparing contingency plans for the agricultural sector.

ABOUT THE AUTHOR

A. S. R. A. S. Sastri received his master's degree in meteorology and oceanography, with specialization in agricultural meteorology, in 1973. He received his doctorate in agroclimatology from Andhra University, Visakhapatnam (India), in 1978. He began his career at the Central Arid Zone Research Institute (CAZRI), Jodhpur (India), in 1975 and remained there until 1982, when he joined the Department of Physics and Agrometeorology at Indira Gandhi Agricultural University in Raipur. In 1989, he was selected by the United States Education Foundation in India (USEFI) for a post-doctoral fellowship under the Fulbright Program. He spent four months (from February to June 1989) at the University of Arizona (Tucson) as a Fulbright post-doctoral fellow, where he worked on rainfall variability and regional climate changes in central India. At present, he is an associate professor in agrometeorology and the head of the Department of Physics and Agrometeorology. Dr. Sastri was awarded a gold medal for team research during 1983-84 by the Indian Council of Agricultural Research (ICAR), New Delhi.

REFERENCES

Azzi, G. 1956. *Agricultural Ecology.* Constable and Co., Ltd., London.

Cocheme, J.; and P. Franquin. 1967. *An Agroclimatology Survey of Semi-Arid Area in Africa, South Sahara.* FAO/WMO Technical Bulletin No. 86. WMO, Geneva.

Das, P. K. 1981. A historical perspective: Droughts and floods in India. *In* R. P. Pearce, ed. *Tropical Droughts: Meteorological Aspects and Implications for Agriculture*; pp. 3-12. WMO, Geneva.

George, C. J.; and Krishna Alde. 1969. *Assessment of Agricultural Droughts from Water Availability Periods.* Pre-Published Scientific Report No. 95. India Meteorological Department, Pune.

George, C. J.; and V. Kalyana Sundaram. 1969. *A Use of Monthly Rainfall Deciles for Assessing Agricultural Droughts in Bihar State.* Pre-Published Scientific Report No. 96. India Meteorological Department, Pune.

Huda, A. K. S.; N. Seetharama; and S. M. Virmani. 1986. Management of agricultural droughts with special reference to soil water availability and water requirements of dryland crops in India. *Jal Vigyan Sameeksha—A Publication of High-Level Technical Committee on Hydrology* 1:53-171.

Krishnan, A.; and K. P. Thanvi. 1971. Occurrence of drought in Rajasthan during 1941-60. *Proceedings of All India Seminar on Dry Farming*, New Delhi.

Landsberg, H. E. 1974. Drought—Summary of lectures and discussion, Geneva, June 1974. *WMO Bulletin* 23(4):226.

Linsley, R. K. Jr.; M. A. Kohler; and J. L. H. Spilham. 1959. *Applied Hydrology.* McGraw Hill, New York.

Patel, S. R.; A. S. R. A. S. Sastri; V. K. Gupta; and B. R. Chandrawanshi. 1986. Crop yields as influenced by agricultural droughts—A water balance approach. *Mausam* 37 (3):341-42.

Penman, H. L. 1948. Natural evaporation from open water, bare soil and grasses. *Proceedings of the Royal Society (A)* 193:120-45.

Ramakrishna, Y. S.; and A. S. R. A. S. Sastri. 1980. Studies on incidence of drought over western Rajasthan. *The National Geographical Journal of India* 26:44-49.

Ramakrishna, Y. S.; A. S. R. A. S. Sastri; G. G. S. N. Rao; and B. V. Ramana Rao. 1984. On prediction of droughts in the Indian arid region. *Mausam* 35:349-54.

Ramana Rao, B. V.; A. S. R. A. S. Sastri; and Y. S. Rama Krishna. 1981. An integrated scheme of drought classification as applicable to the Indian arid region. *IDOJARAS* 85:317-22.

Sastri, A. S. R. A. S.; Y. S. Rama Krishna; and B. V. Ramana Rao. 1981. A new method for classifying agricultural droughts. *Archiv für Meteorologie, Geophysik und Bioklimatologie Ser. B* 29:283-87.

Sastri, A. S. R. A. S.; B. V. Ramana Rao; Y. S. Ramakrishna; and G. G. S. N. Rao. 1982. Agricultural droughts and crops production in the Indian arid zone. *Archiv für Meteorologie, Geophysik und Bioklimatologie Ser. B* 31:405-11.

Sastri, A. S. R. A. S.; Y. S. Rama Krishna; G. G. S. N. Rao; and B. V. Ramana Rao. 1984. Agricultural droughts and crop planning—A case study for western Rajasthan. *Mausam* 35:261-64.

Sastri, A. S. R. A. S.; and R. Malakar. 1981. Climatological analysis of drought over northwest India during the year 1979. *Mausam* 32:259-62.

Sastri, A. S. R. A. S.; and Y. S. Ramakrishna. 1981. Drought during 1979 over western Rajasthan—A climatic perspective. *Annals of Arid Zone* 21:41-47.

Sastri, A. S. R. A. S.; B. V. Ramana Rao; and Y. S. Ramakrishna. 1981c. Climatic analysis of droughts—A case study for 1980 in western Rajasthan. *IDOJARAS* 85:143-48.

Sastri, A. S. R. A. S.; and V. P. Verma. 1985. Water resources in Chhattisgarh region. Unpublished status report. Zonal Agricultural Research Station, Jawaharlal Nehru Agricultural University, Raipur, India.

Sastri, A. S. R. A. S.; and S. R. Patel. 1984. Classification of agricultural droughts of rainfed rice crop—A case study for Central India. *IDOJARAS* 88:223-27.

Sastri, A. S. R. A. S.; and J. S. Urkurkar. 1989. Rainwater management for rainfed rice in Chhattisgarh region of Madhya Pradesh. *Oryza* 26:343-47.

Sastri, A. S. R. A. S. 1988. Agroclimatic features and weather pattern during drought years in Madhya Pradesh with special reference to rice zone. *Seminar on Science and Technology Input and Drought Proofing Strategy*. Central Institute of Agricultural Engineering, Bhopal, India, 17-18 June.

Sastri, A. S. R. A. S.; M. R. H. Siddiqui; and J. S. Urkurkar. 1990. New approaches for drought management of rainfed rice in central India. *Drought Network News* 2(1):12-13.

Sastri, A. S. R. A. S.; and J. L. Chaudhary. 1991. Water balance in bunded rice fields under different rainfed situations in Central India. *International Rice Research Newsletter* 16:28-29.

Subrahmanyam, V. P.; and A. R. Subramaniam. 1964. Application of water balance concepts for a climatic study of droughts in South India. *Indian Journal of Meteorology and Geophysics* 13(37):393-402.

Subrahmanyam, V. P.; and C. V. S. Sastri. 1971. New techniques in drought analysis. *Annals of Arid Zone* 10:120-35.

Thornthwaite, C. W. 1948. An approach towards rational classification of climates. *Geographical Review* 38:55-94.

Thornthwaite, C. W.; and J. R. Mather. 1955. The Water Balance. *Publications in Climatology*, Vol. 8, No. 1. Drexel Institute of Technology, [Philadelphia] Laboratory of Climatology, Centerton, New Jersey, USA.

Wilhite, D. A. 1982. Measuring drought severity and assessing impact. *Proceedings of the International Symposium on Hydrometeorology*; pp. 333-35. American Water Resources Association, Bethesda, Maryland, USA.

Wilhite, D. A.; and M. H. Glantz. 1985. Understanding the drought phenomenon: The role of definitions. *Water International* 10:111-20.

World Meteorological Organization. 1975. *Technical Report on Drought and Agriculture*. Report of the CAgM Working Group on Assessment of Drought; p. 27. WMO, Geneva.

Yao, A. Y. M. 1969. The R-index for plant water requirement. *Agricultural Meteorol*ogy 6:259- 73.

Chapter 6
Planning for Drought: A Methodology

Donald A. Wilhite

INTRODUCTION

Developing a national or provincial drought policy and plan is a complicated but essential first step toward a reduction of societal vulnerability. Until recently, nations had devoted little effort toward drought planning, preferring instead the crisis management approach. Presently, an increasing number of nations are pursuing a more proactive approach that emphasizes the prinçiples of risk management and sustainable development. Because of the multitude of impacts associated with drought and the numerous governmental agencies that have responsibility for some aspect of monitoring, assessment, mitigation, and planning, developing a policy and plan must be an integrated process within and between levels of government. This chapter outlines in considerable detail a generic process that can be adopted by governments that desire to develop a more comprehensive approach to drought management.

DEVELOPING A DROUGHT PLAN: THE SETTING

The factors that may stimulate governments to develop drought plans are numerous and vary from one country to another. These factors may be external, such as the call for the development of drought plans by the World Meteorological Organization (WMO) in 1986 (Obasi, 1986), or internal, such as the occurrence of severe drought and concomitant economic, social, and environmental impacts that significantly affect a nation's economy and progress toward development goals. Although both external and internal factors are important, internal support ultimately must be present for the process to move ahead. Unfortunately, the response efforts of many nations have had little, if any, effect on reducing vulnerability. In fact, vulnerability to drought has increased in some settings because of relief recipients' expectations for assistance from government or donors. If farmers or other relief recipients expect government or donors to assist them during times of distress, this practice will discourage or be a disincentive for self-reliance. In marginal agricultural regions, the provision of relief to farmers may promote land use practices that may not be sustainable in the long term. Disincentives to proper management of the natural resource base characterize the provision of relief in most countries.

The decision to prepare a drought plan almost always rests with a high-ranking political official. This official may initiate the plan development process; if not, the official must be convinced of the need for a plan and the benefits that will accrue if the process is to go forward. This may be a formidable and time-consuming task. Proponents of a plan must begin by determining support for the planning process within key government agencies and assess what expertise exists within the country to assist with the process. Consensus building is an important part of the process that (if done properly) will enhance the chances of successfully initiating and completing the plan. In some cases, a national or regional water resources management or development plan may already exist and a drought plan, once completed, could be incorporated into this broader strategy.

Although the principles of drought planning have been known for some time, progress toward preparedness in most countries has been conspicuously absent. This lack of progress would indicate that impediments or constraints to drought planning exist and must be addressed if the planning process is to be successful.

CONSTRAINTS TO DROUGHT PLANNING

Institutional, political, budgetary, and human resource constraints often make drought planning difficult (Wilhite and Easterling, 1987b). One major constraint that exists worldwide is a lack of understanding of drought by politicians, policy makers, technical staff, and the general public. Lack of communication and cooperation among scientists and inadequate communication between scientists and policy makers on the significance of drought planning also complicate efforts to initiate steps toward preparedness. Because drought occurs infrequently in some regions, governments may ignore the problem or give it low priority. Inadequate financial resources to provide assistance and competing institutional jurisdictions between and within levels of government may also serve to discourage governments from undertaking planning. Other constraints include technological limits such as difficulties in predicting and detecting drought, insufficient data bases, and inappropriate mitigation technologies.

Policy makers and bureaucrats should understand that droughts, like floods, are a normal feature of climate. Their recurrence is inevitable. Drought manifests itself in ways that span the jurisdiction of numerous bureaucratic organizations (e.g., agricultural, water resources, health, and so forth) and levels of government (e.g., national, state, and local). Competing interests, institutional rivalry, and the desire to protect their agency missions (i.e., "turf protection") impede the development of concise drought assessment and response initiatives. To solve these problems, policy makers and bureaucrats, as well as the general public, must be educated about the consequences of drought and the advantages of preparedness. Drought planning requires input by several disciplines, and decision makers must play an integral role in this process.

The development of a drought plan is a positive step that demonstrates governmental concern about the effects of a potentially hazardous and recurring phenomenon. Planning, if undertaken properly and implemented during nondrought periods, can improve governmental ability to respond in a timely and effective manner during periods of crisis. Thus, planning can mitigate and, in some cases, prevent some impacts while reducing physical and emotional hardship. Planning is a dynamic process that must incorporate new technologies and take into consideration socioeconomic, agricultural, technological, and political trends.

It is sometimes difficult to determine the benefits of drought preparedness versus the costs of being unprepared. There is little doubt that preparedness requires financial and human resources that are, at times, scarce. This cost has been and will continue to be an impediment. Preparedness costs are fixed and occur now while drought costs are uncertain and will occur later. Further complicating this issue is the fact that the costs of drought are not solely economic. They must also be stated in terms of human suffering, damage to biological resources, and the degradation of the physical environment, items whose values are inherently difficult to estimate. Post-drought evaluations have shown assessment and response efforts of governments with a low level of preparedness to be largely ineffective, poorly coordinated, untimely, and inefficient in terms of the allocation of resources. Although government expenditures for drought relief are significant and unanticipated, they are usually poorly documented. However, a few examples do exist. During the droughts of the mid-1970s in the United States, specifically 1974, 1976, and 1977, the federal government spent more than $7 billion on drought relief programs (Wilhite et al., 1986). As a result of the drought of 1988, the federal government spent $3.9 billion on drought relief programs and $2.5 billion on farm credit programs (Riebsame et al., 1990). A disaster relief package was also passed by the U.S. Congress in August 1989 in response to a continuation of drought conditions. Between 1970 and 1984, state and federal government in Australia expended more than A$925 million on drought relief under the Natural Disaster Relief Arrangements (Wilhite, 1986). The Republic of South Africa spent R 2.5 billion for drought relief from the mid-1970s to the mid-1980s (Wilhite, 1987). When compared to these expenditures, a small investment in mitigation programs in advance of drought would seem to be a sound economic decision. Congressman George E. Brown, Jr., of California recently suggested that perhaps using as little as one-tenth of one percent of U.S. federal drought relief dollars for preventive measures might lower the costs of future drought relief measures by tens, if not hundreds, of millions of dollars (Brown, 1989). Although this example is from the United States, the principal applies to other political settings. Thus, the rationale for implementing preventive measures must be weighed not only against a retrospective analysis of relief costs but also against future relief costs and savings accrued through reduced economic, social, and environmental impacts. Though difficult to quantify, these savings will be significant.

It is equally important to remind decision makers and policy officials that, in most instances, drought planning efforts will use *existing* political and institutional structures at appropriate levels of government, thus minimizing start-up and maintenance costs. It is also quite likely that some savings may be realized as a result of improved coordination and the elimination of some duplication of effort between agencies or levels of government. Also, plans should be incorporated into general natural disaster and/or water management and development plans wherever possible. This reduces the cost of preparedness substantially. Politicians and many other decision makers simply must be better informed about drought, its impacts, and alternative management approaches and how existing information and technology can be used more effectively to reduce impacts, and at a relatively modest cost.

DEVELOPING A NATIONAL DROUGHT POLICY AND PLAN: A METHODOLOGICAL APPROACH

A planning process was developed recently in the United States to facilitate the preparation of drought plans by state government decision makers (Wilhite, 1990; 1991). This process has been evolving since 1987, when it was first conceived to synthesize the discussions and recommendations of participants of an international symposium and workshop on drought (Wilhite and Easterling, 1987c; 1989). The process was further modified through direct interaction with foreign governments through a series of training seminars on drought management and preparedness. These seminars have been organized and conducted by the International Drought Information Center since 1989.[1] The framework described below presents ten steps considered essential in the planning process (Figure 1). The first four steps actually involve appraising the resources available to support plan development and designing tactics to gain public support for the process. However, the process is intended to be flexible (i.e., governments can add, delete, or modify steps as necessary).

[1] To date, four seminars have been conducted. The first was held in Botswana in September 1989 for countries in the eastern and southern African regions. The second seminar was held in November 1989 in Brazil and focused on the drought-prone northeast region (see Chapter 10). The third seminar was held in Thailand in March 1991 for the Asian and Pacific regions. The most recent seminar was held in Uruguay in March 1993 for the Latin American and Caribbean regions. The primary sponsors for these seminars have been the U.S. National Oceanic and Atmospheric Administration, U.N. Environment Program, World Meteorological Organization, and University of Nebraska. The Organization of American States was a co-sponsor for the Latin American training seminar. Reports of these meetings have been published in *Drought Network News*, the newsletter of the International Drought Information Center, University of Nebraska, Lincoln, Nebraska, U.S.A.

Appointment of
National Drought Commission
(Step 1)

Statement of Drought Policy
and Plan Objectives
(Step 2)

Avoiding and Resolving Conflict
between Environmental and Economic Sectors
(Step 3)

Inventory of Natural, Biological, and Human
Resources and Financial and Legal Constraints
(Step 4)

Development of Drought Plan
(Step 5)

Identification of Research Needs
and Institutional Gaps
(Step 6)

Synthesis of Scientific
and Policy Issues
(Step 7)

Implementation of Drought Plan
(Step 8)

Development of Multilevel Educational
and Training Programs
(Step 9)

Development of Drought Plan
Evaluation Procedures
(Step 10)

Figure 1. The ten-step methodology proposed for the development of a national drought plan (Wilhite, 1991 and 1993).

Step 1: Appointment of National Drought Commission

The planning process is initiated through the appointment of a national drought authority or commission (NDC). The appropriate name for this group (e.g., *commission, committee*, or *task force*) will vary from region to region. The NDC has two purposes. First, during plan development, the NDC will supervise and coordinate the development of the plan. Second, after the plan is implemented and during times of drought when the plan is activated, the NDC will assume the role of policy coordinator, reviewing alternative policy response options and making recommendations to political officials. The NDC is central to this planning process and will be referred to throughout the discussion of the proposed methodology.

The NDC should include representatives of the most relevant mission agencies, recognizing the multidisciplinary nature of drought, its diverse impacts, and the importance of both the assessment and response components in any comprehensive plan, and how this plan must be integrated with long-term development objectives. Agencies to consider for inclusion on the commission are meteorological services, agriculture, water resources, planning, public water supply, natural resources, environmental protection, health, finance, economic and rural development, emergency management, and tourism. A representative from the head of state's office should also be included. Consideration should be given to including key representatives from universities, media (or a public information specialist), and environmental and/or special public interest groups. The purpose of including a representative of the media or a public information specialist is to guarantee that the NDC gives attention to promoting public awareness of drought and associated water issues and the mitigative actions that might be required of government during times of shortage. The actual make-up of this committee would be highly variable from one country to another, reflecting different political infrastructures and the unique combination of economic, social, and environmental impacts associated with drought. Care must be taken to keep the commission membership relatively small so that size does not become in itself a constraint or impediment to the completion of the planning process.

The NDC will need to consider at a later time whether it would be prudent to formalize the plan through the legislative (or some other) process. The danger in not formalizing the plan is that a change in political or administrative leadership may lead to the decay of the plan's infrastructure. It must be emphasized that political interest in drought quickly wanes when the crisis is over, as the hydro-illogical cycle illustrates (Figure 2). Concern and panic during a drought are swiftly replaced by apathy once the rains have returned and drought conditions have abated. Likewise, institutional memory is short. A drought plan (and associated infrastructure) that is ad hoc by nature may cease to exist in a relatively short time. Formalizing the plan after its completion will guarantee that the infrastructure is in place to assist future generations in managing

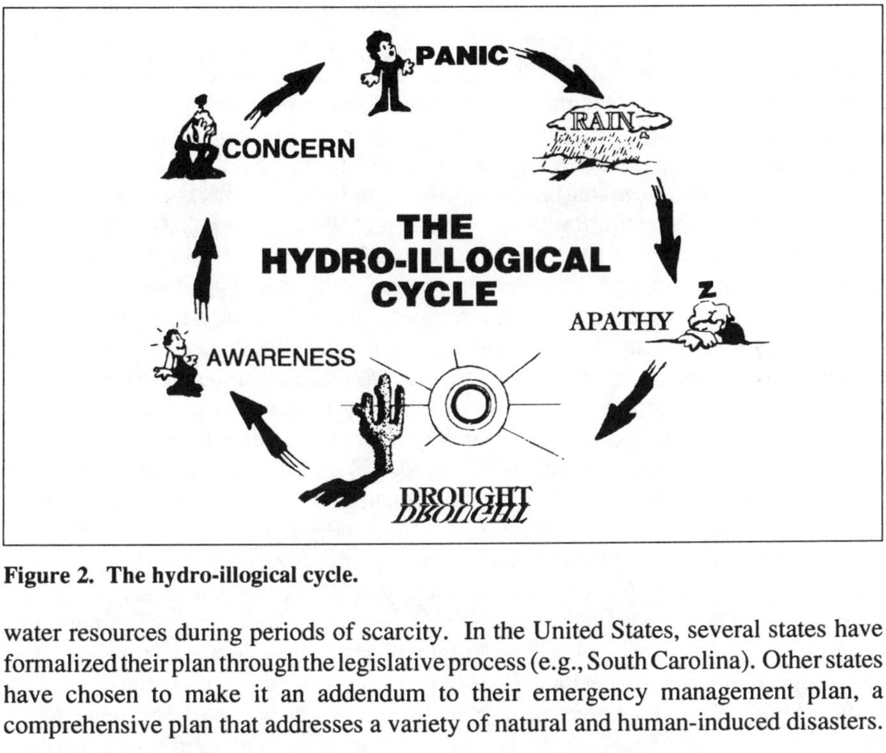

Figure 2. The hydro-illogical cycle.

water resources during periods of scarcity. In the United States, several states have formalized their plan through the legislative process (e.g., South Carolina). Other states have chosen to make it an addendum to their emergency management plan, a comprehensive plan that addresses a variety of natural and human-induced disasters.

Step 2: Statement of Drought Policy and Planning Objectives

As their first official action, the NDC must formulate a national drought policy and the objectives of the drought plan. The objectives of a drought *policy* differ from those of a drought *plan*. A clear distinction of these differences must be made at the outset of the planning process. A drought *policy* will be broadly stated and should express the purpose of government involvement in drought assessment, mitigation, and response programs. Ultimately, the goal of a national policy should be to reduce vulnerability to drought by encouraging sustainable development. Drought *plan* objectives are more specific and action-oriented. Typically, the objectives of drought policy have **not** been stated explicitly by government. What generally exists in many countries is a *de facto policy*, one defined by the most pressing needs of the moment. Ironically, under these circumstances, it is the specific instruments of that policy (such as relief measures) that define the objectives of the policy. Without clearly stated drought policy objectives, the effectiveness of assessment and response activities is difficult to evaluate.

The objectives of drought policy will differ considerably between countries. Based on a comparative analysis of drought assessment and response efforts in the United States and Australia, Wilhite (1986) proposed three objectives of a national policy. First, assistance should encourage or provide incentives for agricultural producers, municipalities, and other water-dependent sectors or groups to adopt appropriate and efficient management practices that help to alleviate the effects of drought. Relief measures relied on in Australia, the United States, and other countries have discouraged self-reliance by encouraging the adoption of management practices that are inappropriate or unsustainable in a particular setting. This objective emphasizes accepting drought as a normal part of climate and preparing for or managing drought risks as a routine course of business. Second, assistance, if provided, should be given in an equitable, consistent, and predictable manner to all without regard to economic circumstances, industry, or geographic region. Assistance can be provided in many forms or as technical aid. Whatever the form, those at risk must know what to expect from government during drought so that they can better prepare to manage that risk. The role of nongovernmental organizations (NGOs) in assistance efforts must also be precisely defined so that these groups complement governmental assistance efforts.

Third, the importance of protecting the natural and agricultural resource base must be recognized. This objective emphasizes the importance of promoting development that is sustainable in the long term. Clearly, many government programs and development projects have been shortsighted, increasing vulnerability to future episodes of drought. For example, agricultural policies that encourage the expansion of agriculture into marginal land areas are not sound when evaluated in the context of sustainability. The development of a national drought policy should lead to an evaluation of all pertinent government programs to ensure that they are not inconsistent with the goals of that policy.

At the initiation of the planning process, members of the NDC should consider many questions pertaining to the development of a national drought policy, including the following:

- What is the purpose and role of government in assessing and responding to drought?
- What should be the scope of the plan (i.e., will it concentrate primarily on agricultural issues or will it be multi-impact in design)?
- What consideration should be given to food supply and distribution or maintaining the nutritional status of various population groups?
- What are the most drought-prone areas of the country?
- What are the most vulnerable sectors of the nation's economy?
- What are the principal social and environmental concerns associated with drought?
- Who are the most vulnerable population groups?

- Will the drought plan be a vehicle to resolve conflict between water users during periods of shortage?
- What resources (human and financial) is the government (and donor organizations) willing to commit to the planning process and in support of the plan once it is completed?
- What are the legal and social implications of the plan?

Following the development of a national drought policy, the next action of the NDC is to identify the specific objectives of the plan. Drought planning is defined as actions taken by individual citizens, industry, government, NGOs, and others in advance of drought for the purpose of mitigating some of the impacts and conflicts associated with its occurrence (Wilhite, 1991). To be successful, drought planning must be integrated between levels of government, involving the private sector, where appropriate, early in the planning process. In the case study section of this book, it will be demonstrated that some governments are now taking a more proactive approach to drought management. For the majority of nations, however, much remains to be done.

A general statement of purpose for a drought plan is to provide government with an effective and systematic means of assessing and responding to drought conditions. Drought plan objectives will, of course, vary between countries, and they should reflect the unique physical, environmental, socioeconomic, and political characteristics of those countries. Objectives that should be considered include the following:

1. To provide timely and systematic data collection, analysis, and dissemination of drought-related information.

2. To establish proper criteria to identify and designate drought-affected areas and to trigger the initiation and termination of various assessment and response activities by governmental agencies, NGOs, and others during drought emergencies.

3. To provide an organizational structure that assures information flow between and within levels of government and defines the duties and responsibilities of all agencies with respect to drought.

4. To develop a set of appropriate emergency and longer-term programs to be used in assessing and responding to periods of water shortage.

5. To provide a mechanism to ensure the timely and accurate assessment of drought impact on agriculture, industry, municipalities, wildlife, health, and other areas as appropriate.

6. To provide accurate and timely information to the media in order to keep the public informed of current conditions and response actions.

7. To establish and pursue a strategy to remove obstacles to the equitable allocation of water during shortages and to provide incentives to encourage water conservation.

8. To establish a set of procedures to evaluate and revise the plan on a continuous basis in order to keep the plan responsive to national needs.

It is suggested that each country consider these objectives and add to, delete, or modify them as appropriate.

Step 3: Resolving Conflict between Environmental and Economic Sectors

Political, social, and economic interests often clash during drought conditions as competition for scarce water resources intensifies, and it may be difficult to achieve compromises under these circumstances. To reduce the risk of conflict between water users during periods of shortage, it is essential for the public to receive a balanced interpretation of changing conditions through the media and from other sources. The NDC should ensure that frequent, thorough, and accurate news releases are issued to explain changing conditions and complex problem areas that exist and situations in which solutions will require compromises on both sides. To lessen the potential for conflict, the views of citizens and environmental and other special interest groups[2] must be considered in the drought planning process at an early stage. Although the level of involvement of these groups will no doubt vary from one setting to another, the power of these interest groups in policy making is worth noting. Public interest organizations in some countries have initiated and participated in the development of natural resource policies and plans for some time and have extensive experience with this process. The involvement of these groups in determining appropriate policy goals strengthens the

[2] These terms are defined according to their meaning in the United States. Other terms may apply in other political settings. The primary difference between special and public interest groups is in what they represent. Where drought is concerned, special interest groups seek to influence policy to benefit their specific economic interests. For example, industry will want to assure sufficient water for corporate operations, and producers of agricultural goods want adequate water supplies for crops and livestock. Public interest groups represent the diverse values of the public domain. During drought, those groups representing natural resources or wildlife interests may be most prominent. Conservation and environmental groups may seek to prevent pollution or disruption of ecosystems. Economic self-interest is not the driving force for these groups. Economics, though, are generally used to define damages resulting from inadequate, inequitable, or inappropriate drought policies. Special and public interest groups may find themselves at odds during a drought crisis. Every effort should be made to incorporate nonlitigious conflict resolution throughout the drought planning process.

overall policy and plan. Moreover, this involvement ensures that the diverse values of society are represented adequately in the policy and plan. Creating an advisory group made up of representatives of these groups is recommended as a means of addressing their concerns.

If it is determined that the public should be involved in drought planning, then that involvement should commence early in the planning process. A drought advisory council (DAC) should be established by the NDC to facilitate this involvement. The DAC should be a permanent feature of the drought plan, assisting the NDC in the flow of information and the resolution of conflicts between water users during severe drought periods.

Public interests and environmental concerns are best addressed early and often in the drought planning process. It is highly desirable to enhance communication between the public and the government at all levels in a drought situation (i.e., assessment, policy formulation, and response effort). The communication networks of public interest and environmental groups can greatly assist government in both dissemination of information and creation of and feedback on mitigation attempts.

Step 4: Inventory of Natural, Biological, and Human Resources and Financial and Legal Constraints

An inventory of natural, biological, and human resources, including the identification of financial and legal constraints, may need to be initiated by the NDC. In most cases, much information already exists concerning available resources, particularly in the natural and biological resource areas. It is also important to determine the vulnerability of these resources to periods of water shortage that result from drought. *Resources* include, for example, physical and biological resources, human expertise, infrastructure, and capital available to government. The most obvious natural resource of importance is water: Where is it located, how accessible is it, of what quality is it? *Biological resources* refer to the quantity and quality of grasslands/rangelands, forests, wildlife, and so forth. *Human resources* include the labor needed to develop water resources, lay pipeline, haul water and hay, process citizen complaints, provide technical assistance, and direct citizens to available services. In addition, representatives of government determine what local, state, or national agencies may be called into action.

Financial constraints would include costs of hauling water or hay, new program or data collection costs, and so forth. These costs must be weighed against the losses that may result in the absence of the drought plan. It should also be recognized that the financial resources available to government vary annually and from one administration to another. This may provide additional incentives for governments to formalize drought plans through the legislative or another process (see Step 1), thus assuring that funds to carry out existing programs are available. *Legal constraints* include user water rights, existing public trust laws, methods available to control usage, requirements for public water suppliers, and emergency and other powers of political and government officials during water shortages.

An inventory of these resources would reveal assets and liabilities that might enhance or inhibit fulfillment of the objectives of the planning process. This systematic survey should include resources available at various levels of government and the often unique resources available at universities. A comprehensive assessment of available resources would provide the information necessary for further action by the NDC. The NDC may also want to undertake an examination of the drought plans available in adjacent and/or climatically similar countries.

Step 5: Development of the Drought Plan

The NDC will be the coordinating body for the development of a drought plan. Once completed, the plan is envisioned to follow a stepwise or phased approach as water conditions deteriorate and more stringent actions are needed. Thresholds must be established such that, when exceeded, certain actions are triggered within government agencies, as defined by the structure of the plan.

A drought plan should have three primary organizational components: monitoring or early warning, assessment of impact, and response. These three organizational components are discussed in detail below. The names given to the components are intended to be generic, principally referring to the function of the committees. Formal linkages will need to be incorporated in the plan for it to function properly and be responsive to provincial and local needs and evolving conditions. An organizational chart illustrating the linkages between the components of the drought plan is shown in Figure 3.

The organizational components shown in Figure 3 represent the recommended structure of a national plan. It is essential that any national plan be integrated with provincial and local levels of government. These linkages are not depicted in the organizational chart. Each of the committees shown in Figure 3 may have a counterpart at the provincial and local level, with well-established linkages to the national committees. These provincial and local committees will facilitate not only data collection and feedback on programs and policies, but also the dissemination of informational products and advisories as well as the implementation of policies.

Monitoring Component: Water Inventory and Outlook Committee (WIOC). A water inventory and outlook committee (WIOC) must be established to monitor current and estimate likely future water availability and moisture conditions. The chairperson of this committee should be a permanent member of the NDC. The WIOC would have five primary duties during the plan development process.

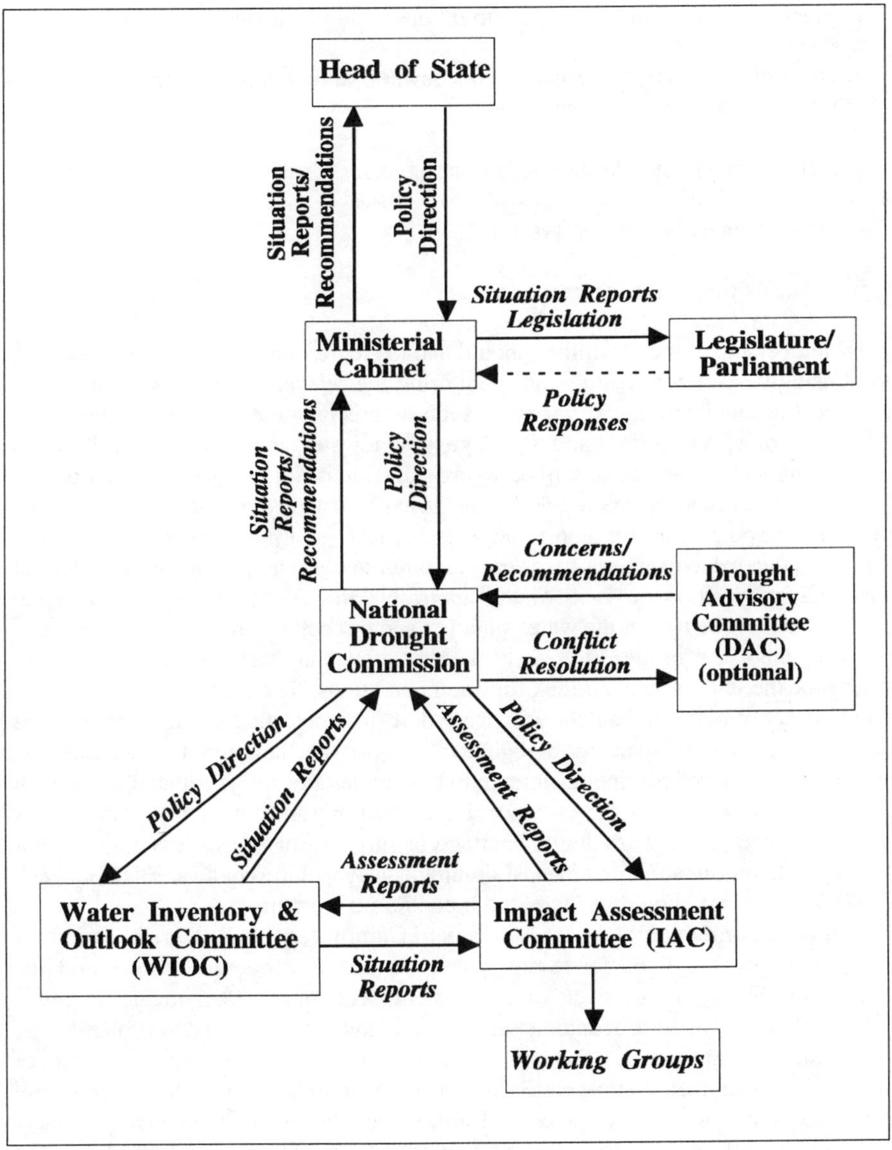

Figure 3. Linkages between the components of the drought plan.

1. Inventory data availability and current observational networks.

2. Determine primary user needs and develop and/or modify current data and information delivery systems.

3. Define drought and develop response triggers.

4. Develop an early warning system.

5. Identify drought management areas.

Membership of the committee should include representatives from agencies with responsibilities for forecasting and monitoring the relevant indicators of the water balance (i.e., meteorological variables such as precipitation and temperature, soil moisture, snowpack, surface water storage, groundwater, and streamflow). Drought monitoring techniques are described in considerable detail in other chapters of this book. In some instances, many agencies at the national and other levels of government may have responsibility for monitoring these indicators. It is not necessary for all of these agencies to have representation on this committee. Rather, it is recommended that data and information on each of the applicable indicators be considered in the committee's evaluation of the water situation and outlook for the country.

It is important for the WIOC to be a permanent committee, meeting regularly to determine the status of and outlook for water conditions. The committee should meet on a monthly basis throughout the year or regularly just preceding and during the period of most concern. One advantage of regular meetings is that the committee will function as a team because of continuous interaction. Another advantage is that a permanent committee can be useful in the early warning of emerging and potentially serious water problems, whether they are due to shortage or surplus situations. It is common for shortage and surplus situations to exist simultaneously within a country. The frequency of WIOC meetings should be increased if conditions warrant.

Impact Component: Impact Assessment Committee (IAC). During periods of drought, impacts will be far-reaching and cut across economic sectors and the responsibilities of various levels of government. The impact assessment committee (IAC) will represent those economic sectors most likely to be affected by drought (e.g., agriculture, transportation). The IAC should be composed of an interagency team of agency heads or their representatives, and its chairperson should be a permanent member of the NDC. It may also be advisable to include university scientists that have expertise in early estimations of impact. The IAC should consider both direct and indirect losses resulting from drought. Often drought assistance is provided only to those experiencing direct losses while agricultural businesses are largely ignored.

Because of the obvious dependency of the IAC on the Water Inventory and Outlook Committee (WIOC), frequent communication between the two is essential.

Two approaches are proposed to assess the magnitude and diversity of impacts that are likely to result from drought. The first approach calls for the IAC to be responsible for determining impacts, drawing information from all available and reliable sources. The advantage of this approach is its simplicity, involving only a select group of agency heads or representatives. This approach will likely be successful in those countries where impacts are concentrated in a relatively few economic sectors (e.g., agriculture). The disadvantage of this approach is that unless an adequate reporting structure is in place to ensure that all impacts are identified and evaluated correctly, indirect effects may go undetected.

The second approach establishes a series of working groups responsible for anticipating and identifying drought-related impacts in all economic sectors deemed to be important. The assessment (and quantification) of drought impacts is complicated and their detection is most difficult without a team of experts from each impact sector working in concert. Members of the IAC may not have the expertise necessary to identify the range of impacts that occur. Each of the working groups would be composed of specialists from various impact sectors. The leader of each working group would be a member of the IAC and would report directly to the IAC. With this model, the responsibility of the IAC is to coordinate the activities of each of the working groups and make recommendations to the NDC. The number of impact sectors or working groups will vary considerably between countries. Working groups used by some states in the United States include municipal water use, wildfire protection, agriculture, industry, commerce, tourism, wildlife, energy, and health. A working group on environmental problems should be considered in most, if not all, instances.

A major point of concern here is that the IAC must give significant attention to the full range of impacts associated with drought and also determine how to target assistance to those economic sectors or vulnerable population groups as the need arises. One of the principal deficiencies of past response efforts has been the inability of government to direct the necessary form of assistance to the economic sector or population group in a timely manner. Assistance that is misdirected or untimely is of little or no value. The IAC must work closely with both the WIOC and the NDC (see next section, Response Component) to ensure that this does not occur.

Response Component: National Drought Commission. The third and final element of a drought plan is the response component. The purpose of this component is to act on the information and recommendations of the IAC and evaluate the range of assistance available from government and other sources to assist agricultural producers, municipalities, and others during times of emergency. Because this is a policy-making body, it would be composed of senior-level policy officials, precisely the same make-up as the NDC. Therefore, in addition to overseeing the plan development process, the NDC should assume the response role following plan development.

During the plan development process, the NDC should inventory all forms of assistance available during severe drought from government and nongovernment sources and evaluate these programs for their ability to address short-term emergency situations as well as long-term mitigation programs to reduce vulnerability to drought. The NDC may want to consider transferring this task to the IAC. The NDC (or IAC) should also recommend other forms of assistance programs that could be developed to respond to drought. During periods of drought, the NDC will make recommendations to the president or appropriate presidential representative concerning specific actions that need to be taken.

Drought assistance should be defined in a very broad way to include all forms of technical and relief programs available from government and nongovernment sources. Rational response options must be determined for each of the principal impact sectors identified by the IAC. These options should examine appropriate drought mitigation measures on three timescales: (1) short-term (reactive or emergency) measures implemented during the occurrence of drought, (2) medium-term (recovery) measures implemented to reduce the length of the post-drought recovery period, and (3) long-term (proactive) measures or programs implemented in an attempt to reduce societal vulnerability to future drought. In many instances, local input should be sought to determine the most rational forms of assistance needed by the various impact sectors.

Societal vulnerability to drought may be influenced substantially by non-drought-related actions taken or policies implemented during nondrought periods. The national drought policy formulated in Step 2 will be especially beneficial at this time. Government must consider the effects of emergency programs on long-term development objectives and guard against implementing emergency programs that draw resources from development programs or interfere with their fulfillment, as has happened in Brazil (see Chapter 10). Emergency programs should foster the achievement of development objectives.

Step 6: Identification of Research Needs and Institutional Gaps

Step 6 is to be carried out concurrently with Step 5. The purpose of this step is to identify research needed in support of the objectives of the drought plan and to recommend research projects to remove deficiencies that may exist. It is unlikely that research needs and institutional gaps will be known until the various committees formed in association with the drought planning process have been through the planning process. Compiling information on research needs and institutional gaps is a function of the NDC. For example, the WIOC may recommend establishing or enhancing an existing groundwater monitoring network. The NDC may find it desirable to create a multidisciplinary scientific advisory panel that could evaluate research proposals, establish funding priorities, and seek financial support from appropriate international or regional organizations, NGOs, or donor governments.

It is likely that institutional deficiencies will be identified as part of Step 6. Agency responsibilities or missions may need to be modified to support activities of the drought plan, modifications that may require legislative action.

Step 7: Synthesis of Scientific and Policy Issues

Previous steps in the planning process have considered scientific and policy issues separately, concentrating largely on assessing the status of the science or on the existing or necessary institutional arrangements to support the plan. An essential aspect of the planning process is the synthesis of the science and policy of drought and drought management. This is the purpose of Step 7.

The policy maker's understanding of the scientific issues and technical constraints involved in addressing problems associated with drought is often negligible. Likewise, scientists generally have a poor understanding of existing policy constraints that affect drought response. A panel of researchers and policy experts recently concluded that communication and understanding between the science and policy communities is poorly developed and must be enhanced if the drought planning process is to be successful (Wilhite and Easterling, 1987a). Direct and extensive contact is required between the two groups in order to distinguish what is feasible from what is desirable for a broad range of science and policy issues. Integration of science and policy during the planning process will also be useful in setting research priorities and synthesizing current understanding. The NDC should consider various alternatives to bring these groups together.

Crucial to this integration process is the provision within the planning process of a means to facilitate scientific information exchange between scientists and policy makers. Since this is not their primary mission, it is unlikely that scientists will freely devote extensive attention to tailoring and otherwise making available research results on a frequent or continuous basis. One way to achieve this interaction is to appoint a specific liaison person or group to facilitate the exchange of information.

Step 8: Implementation of the Drought Plan

The drought plan should be implemented by the NDC to give maximum visibility to the program and credit to the agencies and organizations that have a leadership or supporting role in its operation. As with emergency response plans, all or a portion of the system should be tested under simulated drought conditions before it is implemented. It is also suggested that announcement and implementation occur just before the most drought-sensitive season to take advantage of inherent public interest. In an agricultural setting, this would be in advance of planting or at some other critical time during the growing season. The cooperation of the media is essential to publicizing the plan, and they must be informed fully of the rationale for the plan as well as its purpose, objectives, assessment and response procedures, and organizational framework. If a representative of the media or a public information specialist is a member of the NDC, as recommended, this person should be an invaluable resource in carrying out this step of the planning process.

Training of personnel who will be actively involved in the operation of the plan is also critical if the plan is to achieve its specified goals. This training should include not only persons in the principal national agencies involved in the activated plan, but also persons at the provincial and local levels of government who will provide valuable input into the decision-making process. The key players in the drought plan must thoroughly understand their responsibilities during drought and how these responsibilities relate to those of other organizations and levels of government. If they do not understand the plan and how it functions, it will fail.

In the absence of drought over several consecutive years, the NDC should conduct simulation exercises to keep leadership informed of their responsibilities during drought. This is a common practice in natural disaster mitigation (e.g., earthquakes, hurricanes); it should be no different for drought. Changes in political leadership, retirements, promotions, and transfers to other positions can disrupt the integrity of the plan.

Step 9: Development of Multilevel Education and Training Programs

Educational and training programs should concentrate on several points. First, a greater level of understanding must be established to heighten public awareness of drought and water conservation and the ways in which individual citizens and the public and private sectors can help to mitigate impacts in the short and long term. The educational process might begin with the development of a media awareness program. This program would include provisions to improve the media's understanding of the drought problem and the complexity of the management issues involved, as well as a mechanism to ensure the timely and reliable flow of information to all members of the media (e.g., via news conferences). Second, the NDC should initiate an information program aimed at educating the general population about drought and water management and what they can do as individuals to conserve water in the short run. Educational programs must be long-term in design, concentrating on achieving a better understanding of water conservation issues among all age groups and economic sectors. If such programs are not developed, governmental and public interest in and support for drought planning and water conservation will wane during periods of nondrought conditions.

Step 10: Development of Drought Plan Evaluation Procedures

The final step in the establishment of a drought plan is the creation of a detailed set of procedures to ensure adequate evaluation. To maximize the effectiveness of the plan, two modes of evaluation must be in place:

1. An ongoing or operational evaluation program that considers how societal changes such as new technology, the availability of new research results, legislative action, and changes in political leadership may affect the operation of the plan.

2. A post-drought evaluation program that documents and critically analyzes the assessment and response actions of government, NGOs, and others as appropriate and implements recommendations for improving the system.

The first mode of evaluation is intended to express drought planning as a dynamic process, rather than a discrete event. The operational evaluation program is proposed to keep the drought assessment and response system current and responsive to national needs. Following the initial establishment of the plan, it should be monitored routinely to ensure that societal changes that may affect water supply and/or demand or regulatory practices are considered for incorporation. Accordingly, drought plans should be revised periodically.

The second mode of evaluation is the post-drought audit, which should be conducted or commissioned by governments in response to each major drought episode. Institutional memory fades quickly following drought as a result of changes in political administration, natural attrition of persons in primary leadership positions, and the destruction of critical documentation of events and actions taken. Post-drought evaluations should include an analysis of the physical aspects of the drought: its impacts on soil, groundwater, plants, and animals; its economic and social consequences; and the extent to which predrought planning was useful in mitigating impacts, in facilitating relief or assistance to stricken areas, and in post-drought recovery. Attention must also be directed to situations in which drought-coping mechanisms worked and where societies exhibited resilience; evaluations should not focus only on those situations in which coping mechanisms failed. Provisions must be made to implement the recommendations emanating from this evaluation process. Evaluations of previous responses to severe drought are recommended as a planning aid to determine those actions (both technical and relief) that have been most effective.

The post-drought evaluation review team should address the following questions as a part of the process:

- Was the drought plan followed? If not, why not?
- Were the actions taken and measures implemented effective in mitigating the impact of drought? Which actions and relief measures were effective and which were not?
- Should the plan have included other actions or assistance measures?
- Did aid reach all groups in the stricken area? If not, why not? How were the target groups for aid identified?
- Were the measures timely in relation to the events of the drought period?
- Was it possible to correct errors during the emergency?
- What financial and human resources were allocated to the relief effort? Where did the resources come from and how were they controlled?
- How efficient was the logistical support and the available infrastructure? What obstacles (if any) were encountered that reduced the efficiency of the response?
- How effective was the coordination of response efforts between government, NGOs, and other organizations? How did this cooperation affect the flow of information or assistance?
- Was media coverage accurate and realistic in providing details of the event? What kinds of media were involved? What role did they play in the emergency?

The post-drought evaluation process will identify numerous topics that may require research in order for them to be more adequately addressed during future drought episodes. For example, little is known about the effects of government drought assistance programs. Do they facilitate or hinder the recovery process? Extensive research may be required on the environmental and socioeconomic effects of prolonged rainfall deficiency on various hydrological features such as the depletion of soil water and shallow groundwater. Investigation of the effects of drought on land use, vegetation, and soil is essential to the impact assessment process.

To ensure an unbiased appraisal, governments should place the responsibility for evaluating drought and societal response to it in the hands of nongovernmental organizations such as universities and/or specialized agencies or corporations. An excellent example of this practice in operation is the evaluation of India's Food for Work Program. Although the program is implemented by state government, it is evaluated by an independent body, the Planning Commission (Wilhite and Easterling, 1989). Private foundations, research organizations, and international organizations should be encouraged to support post-drought evaluations.

SUMMARY

A methodology was presented to facilitate the development of a drought plan. This ten-step planning process presents the principal ingredients that should be considered by governments that would like to adopt a more proactive approach to drought management to provide a more effective, efficient, and timely response effort in the short term and reduce societal vulnerability in the long term. Governments are advised to consider this proposed planning process carefully, modifying or adapting it to their particular circumstances by adding or deleting steps as necessary.

The case studies presented in the following chapters of this book reflect many of the considerations described in this chapter that must be addressed in the process of planning for drought. In most settings, planning has been an evolving process, influenced by years of frustration as countries attempted to address drought-related management issues via a reactive, crisis management approach. A growing number of governments have become increasingly aware of the inefficiencies of this approach and are striving to develop sensible national drought policies and plans.

ABOUT THE AUTHOR

Donald A. Wilhite is professor of agricultural climatology in the Department of Agricultural Meteorology and director of the International Drought Information Center at the University of Nebraska-Lincoln. He specializes in studies of the impact of climate on society and societal response to climatic events, particularly drought. Dr. Wilhite is coeditor of *Planning for Drought: Toward a Reduction of Societal Vulnerability*, published by Westview Press in 1987. He has recently written a guidebook on drought preparedness for developing countries under sponsorship of the United Nations Environment Program. Dr. Wilhite is chair of the Committee on Applied Climatology of the American Meteorological Society.

REFERENCES

Brown, G. E. 1989. *National Agricultural Weather Information System Act.* Hearing before the Subcommittee on Department Operations, Research, and Foreign Agriculture of the Committee on Agriculture, U.S. House of Representatives on H.R. 1880. 101st Congress, 3 May.

Obasi, G. P. 1986. Drought Response Plans. Memo from the Secretary-General of WMO to Permanent Representatives of Members of WMO, 14 May. WMO, Geneva.

Riebsame, W. E.; S. A. Changnon, Jr.; and T. R. Karl. 1990. *Drought and Natural Resources Management in the United States: Impacts and Implications of the 1987-89 Drought*. Westview Press, Boulder, Colorado, U.S.A.

Wilhite, D. A.; N. J. Rosenberg; and M. H. Glantz. 1986. Improving federal response to drought. *Journal of Climatology and Applied Meteorology* 25:332-42.

Wilhite, D. A. 1986. Drought policy in the U.S. and Australia: A comparative analysis. *Water Resources Bulletin* 22:425-38.

Wilhite, D. A.; and W. E. Easterling, eds. 1987a. *Planning for Drought: Toward a Reduction of Societal Vulnerability.* Westview Press, Boulder, Colorado, U.S.A.

Wilhite, D. A.; and W. E. Easterling. 1987c. Drought policy: Toward a plan of action. *In* D. A. Wilhite and W. E. Easterling, eds. *Planning for Drought: Toward a Reduction of Societal Vulnerability*; Chapter 37. Westview Press, Boulder, Colorado, U.S.A.

Wilhite, D. A.; and W. E. Easterling. 1987b. Introduction (Workshop Summary). *In* D. A. Wilhite and W. E. Easterling, eds. *Planning for Drought: Toward a Reduction of Societal Vulnerability*; Chapter 34. Westview Press, Boulder, Colorado, U.S.A.

Wilhite, D. A.; and W. E. Easterling. 1989. Coping with drought: Toward a plan of action. *Eos* 70(7):97, 106-8.

Wilhite, D. A. 1987. The role of government in planning for drought: Where do we go from here? *In* D. A. Wilhite and W. E. Easterling, eds. *Planning for Drought: Toward a Reduction of Societal Vulnerability*; Chapter 25. Westview Press, Boulder, Colorado, U.S.A.

Wilhite, D. A. 1990. *Planning for Drought: A Process for State Government.* IDIC Technical Report Series 90-1, University of Nebraska, Lincoln, Nebraska, U.S.A.

Wilhite, D. A. 1991. Drought planning: A process for state government. *Water Resources Bulletin* 27(1):29-38.

Wilhite, D. A. 1993. *Preparing for Drought: A Guidebook for Developing Countries.* United Nations Environmental Program, Nairobi, Kenya (in press).

Chapter 7
Implications of Global Warming for Climate Variability and the Occurrence of Extreme Climate Events

Linda O. Mearns

INTRODUCTION

Drought is undoubtedly the best-studied extreme climatic event. As an extreme event, changes in its frequency for any given location are affected by changes in mean conditions of climate (including, among other climate variables, temperature and precipitation) as well as the variability (on different time scales) of climate and other higher-order statistical moments (Mearns et al., 1984; Wigley, 1985; Wigley, 1988). It has recently been demonstrated (Katz and Brown, 1992) that changes in the frequencies of extreme events (e.g., the occurrence of maximum temperatures above 35°C) are more sensitive to variability (i.e., standard deviation) changes than to mean changes in a climate element. This relative high sensitivity to variability should also obtain for more complex extreme events that are functions of several climate variables. Drought is an unusually complex extreme event. It is something of a creeping phenomenon; neither its onset nor its end is clearly punctuated in time. It is difficult to measure drought severity, since drought is a combination of factors such as duration, intensity, and areal extent. In addition, different types of drought are recognized, such as meteorological drought, agricultural drought, and hydrologic drought.

Although drought is not the easiest extreme event to define and quantify, it is one of the most important extremes to be considered in evaluating changes in climate related to possible global warming from the continued buildup of greenhouse gases. The focus of this chapter is the discussion of drought in the context of global climate change, particularly how its incidence and severity will be affected by changes in the mean and variability of climate elements, chiefly temperature and precipitation. The importance of changes in climatic variability in particular will be emphasized where possible since this has been a much less studied phenomenon despite its great importance. Also included is a brief discussion of some recent work on the possible agricultural impacts of changed climate variability as it pertains to drought.

Author	Year	Region	Variable	Results
Van Loon & Williams	1978	N. America	Interannual T	No relation
Chico & Sellars	1979	United States	Interannual T not studied	Trend, but relation
Diaz & Quayle	1980	United States	Interannual T&P	No relation
Brinkmann	1983	Wisconsin	Daily T	Daily variability and mean
Lough et al.	1983	N. Hemisphere	Seasonal T&P	Limited relation
Schuurmans & Coops	1984	Europe	Seasonal T	Winter high latitudes
Skeeter et al.	1988	United States	Daily T	Limited relation

T = Temperature P = Precipitation Results = whether a relation between the mean and variance of the variable over time was found

Table 1. Studies of relations between mean and variance changes of climate.

VARIABILITY, GLOBAL WARMING, AND DROUGHT

Given the importance of changes in climate variability in altering the frequency of extreme events such as drought (Katz and Brown, 1992), investigation of how variability may change with global warming becomes a particularly important research task. Efforts to gain insight into possible relationships between changes in mean climate and variability have focused both on relations in the historical past and in climate simulation experiments.

Empirical Studies—Detecting Changes in Variability

One of the methods available for possibly gaining some insight into how climate variability may change in a generally warmer climate is to investigate past relationships in the climate record between mean climate change and changes in variability.

Temperature. Studies attempting to find relationships between changes in mean temperature and the variability of temperature have now been reviewed a number of times (e.g., Carter and Parry, 1986; Mearns, 1989 and 1992a). Table 1 summarizes the results of major studies for both temperature and precipitation. In summary, the studies have failed to find distinct correlations, and in particular they have failed to confirm the hypothesis put forward by some researchers (e.g., Bryson, 1975; Tyson, 1977) that temperature variability should decrease with general mean global warming conditions. In fact, the study by Diaz and Quayle (1980) indicates a slight tendency in the conterminous United States for warmer periods in the twentieth century to have higher variability of annual temperature than cold periods.

Precipitation. As Table 1 indicates, very few studies have been made of the relation between changes in mean precipitation and variability of precipitation. Yet, obviously, numerous studies (Diaz, 1983; Carter and Parry, 1986; Barnett, 1985; Bradley et al., 1987) have documented temporal variability of precipitation related to drought. It is clear that precipitation amounts do significantly vary. What is completely unclear is whether there is any relation between mean changes and variability of precipitation. Very few studies trying to demonstrate such a relation have been produced because results have been highly variable. Diaz and Quayle (1980) and Lough et al. (1983) found generally no relation between changes in mean precipitation for a series of years and changes in precipitation variability, or, more specifically, they found a great spatial variability of relations and variations according to season.

Difficulties in Variability Change Detection. Aside from the spatially discontinuous nature of precipitation, and its high spatial variability, other characteristics make even determining the existence of a difference in variability difficult. Statistical tests designed for measuring changes in variability (i.e., variance), such as the F-test, are very sensitive to assumptions of normality, and precipitation data, even aggregated annually, tend to be highly skewed. In addition, the variance of precipitation time series is not the best measure of variability, again because of the non-Gaussian character of precipitation distributions. Hence, other statistical tests are preferable (such as tests of difference between interquartile ranges [Mearns et al., 1990]). In the case of statistical tests for variability, a larger sample size is often required to render the tests sufficiently robust (Katz, 1992).

GCM Studies, Climate Variability, and Drought

Current Knowledge of Climate Change. Most of our knowledge regarding climate change comes from the results of experiments with general circulation models (GCMs), which simulate the behavior of the climate system. What is currently believed likely regarding climate change, based on climate model results of doubled CO_2 experiments, is the following (Houghton et al., 1990 and 1992): (1) increase in surface air temperature on a global scale between 1.5 and 4.5°C, with greater warming in high latitudes and during the winter; (2) global mean increase in precipitation, but significant increases and decreases for particular regions and seasons. In response to doubled CO_2, most models show increased precipitation in high latitudes and the tropics throughout the year and in winter in mid-latitudes (Mitchell et al., 1990).

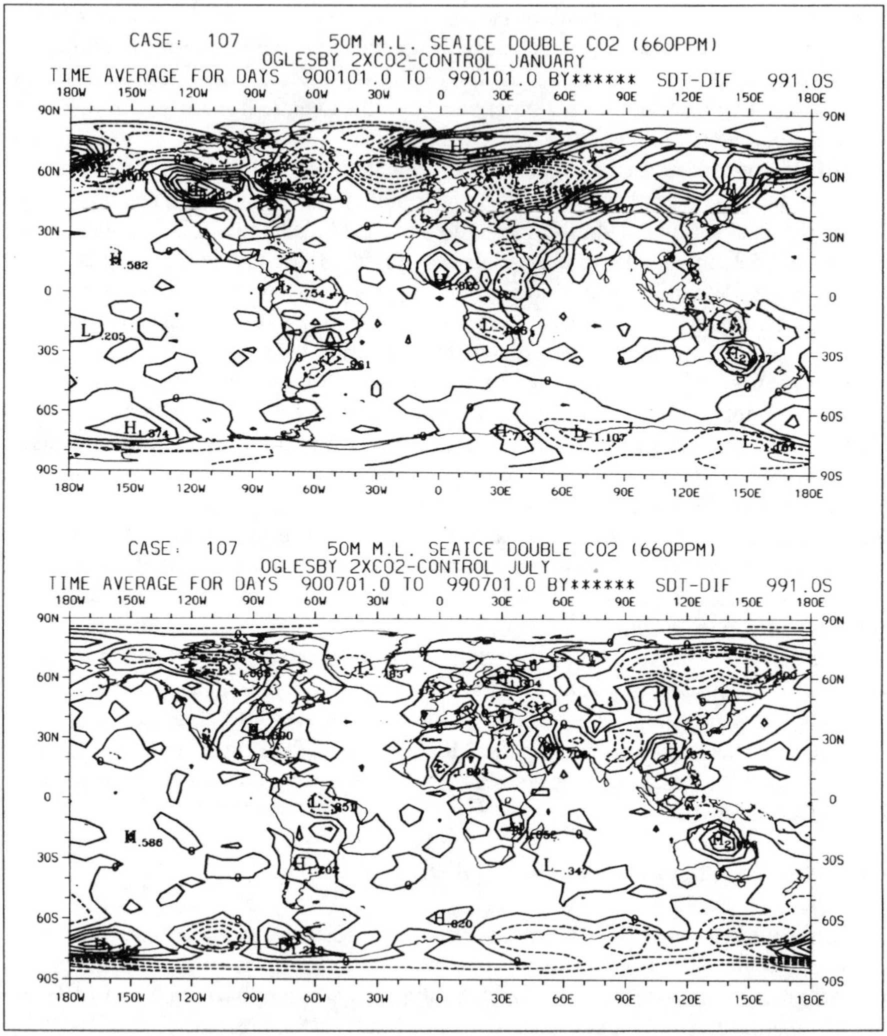

Figure 1. Difference in standard deviation ($2xCO_2$ minus control) for mean January (a) and July (b) temperature for runs of the CCM1 by Oglesby and Saltzman (1990). Dashed contours indicate negative values, solid contours positive values. Contour interval is 0.5°C.

Even less is known about possible changes in variability of climate from daily to interannual to decadal time scales. Although arguments regarding decreases in the variability of temperature under conditions of global warming have been put forward (Rind, 1986 and 1991), most recent analyses (Cao et al., 1992; Whetton, personal communication; Mearns, 1992a) indicate regional variations in the direction of change of temperature variability. In this regard, modeling results support the empirical study results presented earlier. Figure 1 presents an example of changes in interannual variability of temperature (difference in standard deviations calculated by Mearns, 1992b) of the doubled CO_2 climate minus the control run for climate experiments performed by Oglesby and Saltzman (1990). In both January and July (a and b), significant increases and decreases in variability change are indicated. For example, in the Great Plains of North America, most of the region shows increased variability of temperature under doubled CO_2 conditions. And these conditions are similar for every month of the year. These results should not be taken as any more valid than others. However, they are illustrative of most recent results from equilibrium doubled CO_2 experiments using mixed layer oceans.

Interestingly, more of a consensus exists regarding changes in precipitation variability, although this is partially a result of its inherent high spatial variability. There seems to be some indication that variability of precipitation will increase, at least at locations and during seasons when mean precipitation increases. This is partially based on the correlation in the results of Rind et al. (1989), where for 70% of the climate model grid boxes, mean precipitation increases result in increased absolute precipitation variability. This tends to be the case in other regional study results (Wilson and Mitchell, 1987; Mearns et al., 1990). Returning to the analysis of changes in variability of the runs of Oglesby and Saltzman (1990), the correlation between direction of precipitation change and direction of variability change is obvious from visual inspection alone of Figure 2, which shows changes in mean precipitation and precipitation variability for June.

There does seem to be something of a consensus that with global warming, convective precipitation would increase in general at the expense of large-scale circulation precipitation; hence, greater intensity of precipitation events could be anticipated (Mitchell et al., 1990). This is born out by the results of Gordon et al. (1992) and Pittock et al. (1991) for the CSIRO climate model. The likelihood of increased local convective instability is less dependent on the particular climate model used, since the radiative heating and cooling characteristics of the models are similar (Mitchell et al., 1987).

Simulation of Drought. A number of GCM experiments have been conducted concerning the development and causes of regional drought conditions. These typically involve testing a hypothesis to determine if and how a particular kind of external forcing to the climate system results in drought conditions in a particular area. Typical

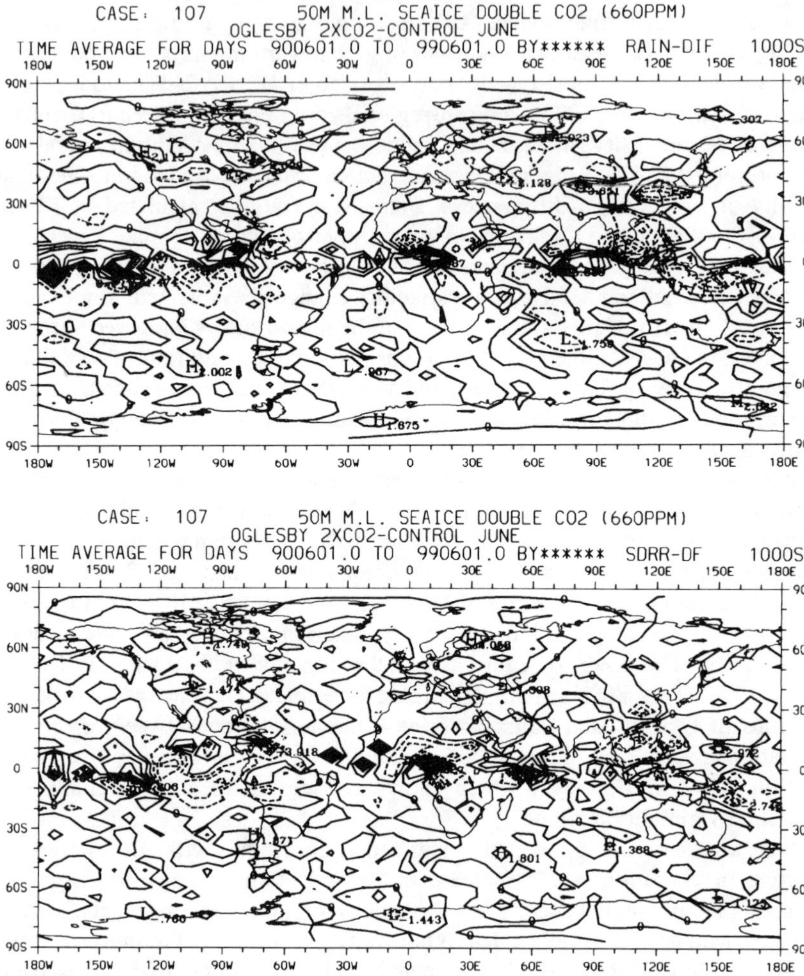

Figure 2. Difference in June precipitation (2xCO_2 minus control runs) of the CCM1 by Oglesby and Saltzman (1990), mean (a) and standard deviation (b). Dashed contours indicate negative values, solid contours positive values. Contour interval is 1.0 mm/day.

examples include the effect of sea surface temperature (SST) anomalies on drought in the Sahel (Druyan, 1988; Druyan and Hastenrath, 1991), feedbacks between initial soil moisture anomalies and prolongation of drought conditions (Oglesby and Erikson, 1989), and fluctuations in the strength of the Asian monsoon (Palmer et al., 1992; Kitoh, 1992). Some of the studies have been quite successful at determining the causes of drought conditions in a particular year. The experiment of Trenberth et al. (1988) provides a compelling example. They demonstrated that the 1988 North American drought could be successfully reproduced by forcing a general circulation model with the observed anomalous SSTs in the tropical eastern Pacific for that year (SSTs in the tropical Pacific were colder than average).

Particular focus on changes in drought frequency per se in GCM climate change studies has been relatively recent, and studies focusing explicitly on changes in climate variability and drought incidence have been all but nonexistent. However, as will be discussed in the following section, a number of studies dealing with changes in interannual variability of climate implicitly concern themselves with drought incidence.

The most complete study to date (Rind et al., 1990) concerns analysis of the likelihood of future drought based on the results of climate change experiments with the Goddard Institute for Space Studies (GISS) GCM. Year-to-year climate variability found in the climate change experiment (Rind et al., 1989) is taken into account in part of the study. Essentially drought, defined according to the Palmer Drought Severity Index (PDSI) and the "supply-demand" drought index [SDDI] (based on the difference between potential evapotranspiration and precipitation), is analyzed in the transient run of the GISS model (Hansen et al., 1988), wherein the time evolving response of climate to incrementally increasing greenhouse gases is simulated. Calculations are also performed for an equilibrium doubled CO_2 run. Of particular importance is the finding that the increased potential evapotranspiration in the increased CO_2 cases, which is largely a function of increasing surface air temperature, is the factor that drives the increasing occurrence of drought based on the SDDI or the PDSI. Changes in precipitation are of secondary importance. According to these indices, drought becomes more prevalent and severe in the course of the transient run. For example, drought that according to the PDSI occurred in the United States 5% of the time during the summer under current climate conditions would occur around 40% of the time by the 2050s, according to the transient run results for that future decade.

Although earlier studies documented predicted change in soil moisture under doubled CO_2 conditions from GCM results (Manabe and Wetherald, 1987; Kellogg and Zhao, 1988; Zhao and Kellogg, 1988), these studies probably have underestimated the occurrence of drought because of the relatively crude representation of soil moisture in the GCMs studied. Rind et al. (1990) demonstrate the limitations of analyzing soil moisture changes in GCMs with very crude soil moisture calculations, based on the very unrealistic evaporation efficiencies used. It is important to note in the result of Rind et al. (1990) that often on the basis of soil moisture changes calculated in the model runs, serious drought was not indicated as it was by the drought indices.

Oglesby (1991) recently examined the role of natural climatic variability in producing North American drought conditions using the NCAR Community Climate Model. Relative decreases in soil moisture compared to the mean in the model are considered equivalent to drought, with no specific criterion used for determining a soil moisture threshold that hallmarks a drought. By holding external boundary conditions constant (i.e., having only seasonally varying sea surface temperatures), he determined that the natural climatic variability of the atmospheric model was able to produce reduced soil moisture conditions in the absence of any external forcing mechanism such as sea surface temperature anomalies that occur in El Niño/Southern Oscillation (ENSO) events (see below). This point is important since it implies that drought occurs naturally in the climate system even without fluctuations in boundary conditions such as sea surface temperatures.

The most complete modeling experiments to date are those of coupled atmosphere/ ocean models (Washington and Meehl, 1989; Cubasch, 1991; Stouffer et al., 1989). Some of these model experiments reveal differences in soil moisture reductions compared to those of equilibrium response experiments with simple mixed layer ocean models described above. For example, Manabe et al. (1992) determined in their transient run with a coupled model that, because of the thermal inertia of the ocean, the warming of the ocean surface is delayed, the moisture supply from the oceans to the continents is thereby reduced compared to the mixed layer experiment, and the summer reduction of soil moisture in the mid- to high latitudes of the Northern Hemisphere seen in many equilibrium experiments is enhanced. These results could indicate that analyses of drought from modeling experiments using simple mixed layer models of the ocean underestimate the severity and occurrence of the phenomenon.

Simulation of Precipitation. The importance of changes in potential evapotranspiration in the determination of the possible future occurrence of drought is emphasized in Rind et al. (1990), given that precipitation is much more poorly represented in climate models than is temperature (Boer et al., 1991 and 1992). Since potential evapotranspiration is largely determined by temperature, changes in drought occurrence in models will be less dependent on changes in precipitation than changes of temperature. There is greater confidence in predictions of changes in surface air temperature than in changes in precipitation, both on a zonal and regional basis.

Figure 3 shows latitudinal zonal average values of precipitation for observed and model results for 14 different GCMs (Boer et al., 1991). Although the overall patterns are similar in the observed and model precipitation values, serious over- and underestimations exist in the absolute values. Most models overestimate precipitation in polar regions and northern hemisphere mid-latitude winters and underestimate it in southern

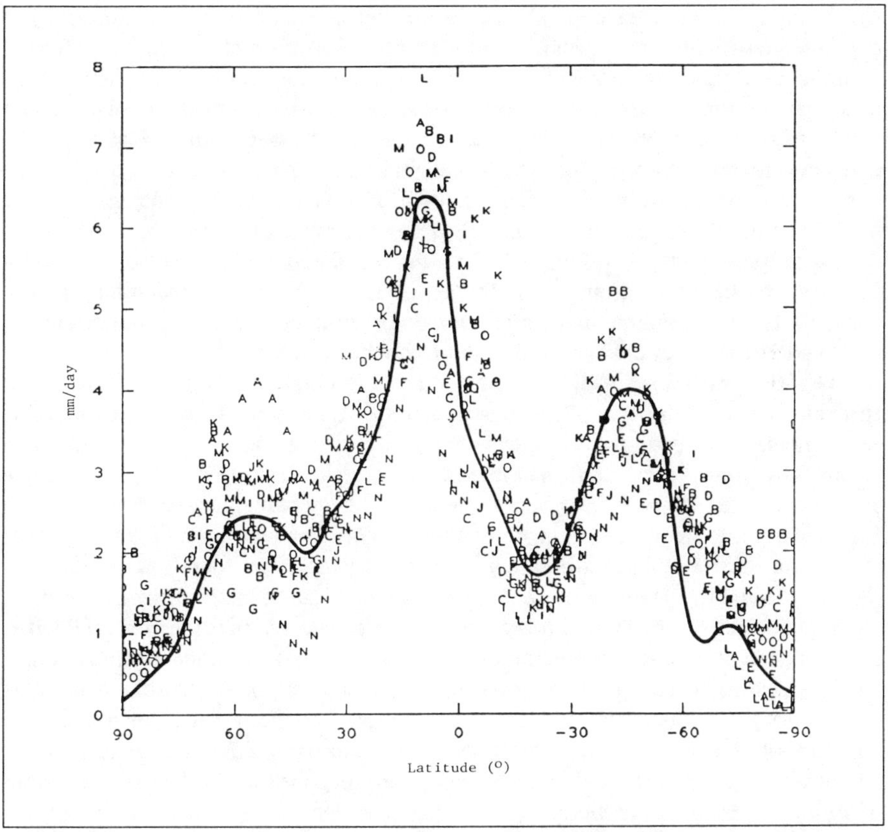

Figure 3. Comparison of zonally averaged GCM control run simulations of precipitation (mm/day) for 14 models (indicated by letters) compared to observations (solid line) for June, July, August. (Sample models: A=MRI [Meteorological Research Institute, Japan]; B=BMRC [Bureau of Meteorological Research Centre, Australia]; C=GFDL [Geophysical Fluid Dynamics Lab]; D=NCAR low resolution; F,J=CCC [Canadian Climate Centre]; L=UKMO [United Kingdom Meteorological Office].) Source: Boer et al., 1991.

hemisphere mid-latitude summer. These errors become much more pronounced when examined on a regional basis (Mearns et al., 1990; Grotch, 1988). A further complication exists regarding validation of precipitation in GCMs in that greater disagreement appears among different observational data sets, compared to data sets for temperature (Hulme, 1991). This further complicates the determination of a model's adequacy to reproduce the observed precipitation climatology. Obviously if no agreement exists regarding observed precipitation, then it is harder to determine if the model is properly simulating this unknown observed quantity. Although changes in evapotranspiration driven primarily by changes in temperature may be most important for average change in drought conditions, certainly changes in precipitation, and particularly in the interannual variability of precipitation, are very important for refining knowledge of changes in drought frequency and severity.

GCM Simulation of El Niño Southern Oscillation Events—Changes in Interannual Variability. The limited ability of most of the climate modeling experiments described above to clearly indicate how the frequency of drought may change is partially a function of limited ability to successfully model some ocean/atmosphere interactions that account for much of the interannual variability of climate (particularly of precipitation) and result in significant variations in sea surface temperatures (SSTs). The most extreme variations of SSTs are found as part of ENSO (El Niño Southern Oscillation) events. ENSO, which can be characterized roughly as an irregular oscillation of the coupled atmosphere/ocean system in the tropical Pacific, is associated with significant precipitation anomalies in the tropics and beyond (Kiladis and Diaz, 1989), which have had serious impacts on human society worldwide (Glantz et al., 1987). Because ENSO events are the most important source of interannual climatic variability, how well climate models reproduce the occurrence of such events and how well they appear to "predict" changes in this phenomenon under conditions of global warming is particularly important for determining changes in drought frequency and intensity under conditions of global warming. In considering this phenomenon, the adequacy of different types of climate models for reproducing such events becomes quite important.

So far, most of the climate models discussed in earlier sections (except Manabe et al., 1992) were those which incorporate a very simple mixed layer ocean, usually with a 50 m depth, which has a heat capacity but does not model the dynamics of a three-dimensional ocean. Although such models do exhibit some interannual variations of SSTs (Meehl, 1990), they are not adequate for the explicit prediction of changes in oceans relevant to the reproduction of ENSO events.

However, a number of atmospheric model tests do indicate that the atmospheric models, when provided with SST anomalies representative of ENSO events, do respond in ways similar to observed atmospheric pattern responses to ENSO events (Pitcher et al., 1988). For example, a recent study (Gordon and Hunt, 1991) using the CSIRO four-layer GCM indicated that simulations of rainfall perturbations forced by sea-surface temperature anomalies in the Pacific ocean representing El Niño and anti-El Niño events were successful. Droughts in eastern Australia and Venezuela were generated for the El Niño conditions, as well as flooding conditions in Peru.

The success of these models in responding to sea-surface temperature anomalies is certainly heartening; however, under conditions of global warming, the ability to predict changes in SSTs and related characteristics of ENSO events becomes paramount. For this purpose, full coupled atmospheric/oceanic GCMs are required. The experiments of Philander et al. (1992) and Latif et al. (1991) indicate that coupled models can successfully simulate interannual variations that are somewhat consistent with aspects of ENSO events, although the amplitude of the variations tends to be underpredicted. Meehl (1990) indicates that the Washington and Meehl (1989) coupled climate run simulated some aspects of ENSO events, and Meehl et al. (1992) went on to investigate changes in ENSO events under global warming conditions. They found that the patterns of tropical precipitation and soil-moisture anomalies are similar in control and doubled CO_2 cases, but dry areas became drier and wet areas wetter (i.e., variability increases). These preliminary results indicate that increased variability of precipitation engendered by changes in ENSO events could lead to an even higher incidence of drought compared to a consideration of only mean changes in temperature and precipitation. However, confirmation of such changes with other coupled GCM experiments remains to be achieved.

IMPACT STUDIES RELATED TO CHANGES IN TEMPERATURE AND PRECIPITATION VARIABILITY

The above discussions still make it less than clear, in a detailed quantitative sense, how climate variability changes can increase the frequency and intensity of drought. Given the difficulties of clearly defining drought for all contexts, this lack of clarity is not surprising. One way of demonstrating how climate variability changes can alter the frequency of drought, however, is to provide a context-specific analysis of the effect of changes in the major climate variables involved in drought (temperature and precipitation) in a specific impacts context.

One of the most common impact areas concerned with fluctuations in climate related to drought is that of agriculture (Parry and Carter, 1985). Numerous studies have addressed the question of how changes in climate could affect agricultural production (e.g., Rosenzweig, 1990; Smith and Tirpak, 1989), but few have concerned themselves with changes in climatic variability except in an implicit sense when using climate change analogues (e.g., Brklacich and Smit, 1992; Easterling et al., 1992). In this kind of study, some period in the historical past is used as an analogue to climate change (such as the 1930s in the United States), and if the variability of climate was different from

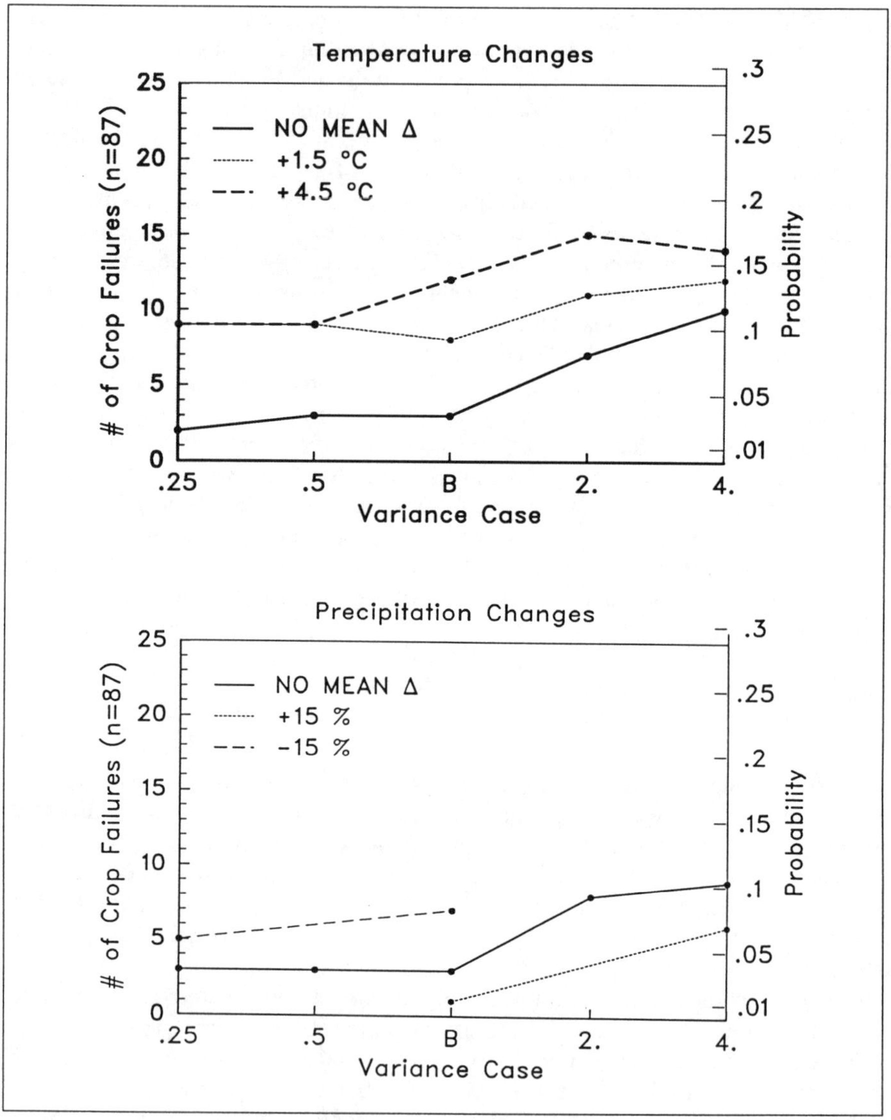

Figure 4. Probability of fallow crop failure at Goodland, Kansas, for different mean and variance changes of (a) temperature and (b) precipitation simulated by the CERES-wheat model. Source: Mearns et al., 1992a.

the present for the analogue, then the study implicitly concerns itself with the different variability. Recently, a few studies have explicitly dealt with changes in variability, which have direct bearing on drought occurrence and its effects (Panturat and Eddy, 1989; Mearns et al., 1992a, Wilks, 1992). Recent work by Mearns et al. (1992a and 1992b) provides such an example. In this study, the effect of changes in mean temperature and precipitation and in their interannual variabilities on winter wheat crop production in the central Great Plains of the United States is analyzed.

Using the CERES-Wheat deterministic crop model, the authors investigate changes in the probabilities of crop failures as a function of changes in mean and interannual variability of climate. Since two of the three types of crop failure simulated by the wheat model result directly from significant moisture deficits, these types of crop failures can be viewed as surrogates for agricultural droughts. Figures 4 a and b show results for changes in temperature and precipitation, respectively, for one location in western Kansas. In both cases, increasing variability of climate results in increased crop failures. Most of the increases in crop failures are caused by increased soil moisture deficits from anthesis to beginning of grainfill, which results in a failure to attain the criterion for minimum number of grain kernels/m^2. This is the most prevalent kind of crop failure. Note particularly in Figure 4b that, even with a mean precipitation increase of 15%, an increase in precipitation variability (i.e., doubling the standard deviation of monthly mean precipitation amounts) still increases crop failures due to moisture stress, and there are more crop failures than in the base case (B, no change in mean or variance). Note also that a decrease in precipitation variability does not result in a proportionate decrease in crop failures, only a very slight one.

In an additional part of this study (Mearns et al., 1992b), combinations of precipitation and temperature variability changes were applied. In the case of increased variability of temperature and precipitation (with no mean changes), the highest number of crop failures were attained (from a combination of winter kill and moisture stress). If one accepts the operational definition of agricultural drought in this study, then the clear exacerbation of drought conditions for wheat production as a result of changes in the interannual variability of temperature and precipitation is demonstrated.

FUTURE RESEARCH NEEDS

Improvements in GCMs

The most important research need related to improved analysis of possible changes in the occurrence and severity of drought concerns key improvements in general circulation models. Most of the research areas for GCMs described here are well underway, but none have developed sufficiently to suggest that further development is unnecessary. The three key areas in which improvements in the modeling of physical processes are being made are modeling of surface processes, modeling of oceans and coupling of ocean and atmosphere models, and modeling of clouds. A further issue is the improvement of the horizontal and vertical spatial resolution in the climate model, which would in and of itself improve the modeling of many atmospheric and oceanic processes. All of these improvements are important for developing reasonably accurate regional representation of climate (Dickinson, 1989), but some have particular implications for simulating drought.

Until relatively recently, most GCMs used rather simple surface process parameterizations. Very simple surface albedo schemes were used, vegetative effects in general were not included, and the surface hydrology was of the "bucket" method. It is this kind of oversimplification of the calculation of soil moisture that Rind et al. (1990) criticized in their discussion of drought prediction by GCMs. In the past five years or so, a number of sophisticated surface packages have been developed, which explicitly model vegetative effects on the surface energy balance and use much more detailed soil moisture parameterizations (Dickinson and Henderson-Sellers, 1988; Sellers et al., 1986; Verseghy, 1991). Among other things, these improvements will render the soil moisture changes predicted in climate change experiments more useful for analyzing drought. Although these models have been available now for more than a few years, many climate change experiments are still run with models possessing very simple surface schemes (e.g., Oglesby, 1991; Washington and Meehl, 1989).

The importance of developing coupled ocean/atmosphere models has already been discussed in the section on ENSO events. To properly simulate the time-evolving response of the earth/atmosphere system to slowly increasing greenhouse gases and particularly to simulate interannual variability adequately, these types of models are essential. Although a number of experiments have already been run (Washington and Meehl, 1989; Cubasch, 1991; Manabe et al., 1992), the models are still relatively crude, and serious difficulties in coupling the models still remain (Bretherton et al., 1990). Further development of these models (i.e., improving the spatial resolutions of both ocean and atmospheric models), including solving the flux adjustment problem at the interface between the models, currently is one of the most significant research areas for climate modeling.

Proper modeling of the role of clouds in the radiation balance of the earth/atmosphere system has proven to be a difficult task. Most clouds are sub-grid scale (i.e., not explicitly resolved in the climate model) and hence their rather complex properties and interactions must be highly parameterized. The modeling of clouds has received much attention recently (Mitchell et al., 1989; Slingo and Slingo, 1991), and more explicit models are being developed. Also, this problem is being alleviated somewhat by improvements in horizontal spatial resolution. However, in regard to precipitation, improved resolution is not necessarily sufficient. Boer et al. (1991 and 1992) found that the higher resolution models they examined had not significantly improved on earlier models in their ability to accurately reproduce zonal precipitation.

Although improvement in resolution per se has not solved the cloud radiative feedback problem, it is still believed that higher spatial resolution will improve the representation of many processes. Experiments by Boer and Lazare (1988), Kiehl and Williamson (1991), and Boville (1991) indicate significant improvements in the simulation of climate with higher spatial resolution. However, improvements in resolution are partially hampered by available computer time for and the expense of running climate models (Giorgi and Mearns, 1991).

Another approach to the resolution problem is to embed very high resolution (60 km) mesoscale models within GCMs to simulate one region with great detail (Giorgi and Mearns, 1991). This technique has already been used to generate climate change scenarios for Europe (Giorgi et al., 1992a) and most recently for North America (Giorgi et al., 1992b).

Adequacy of Impact Models

Although improvements in GCMs will be the most important area of research for predicting changes in drought frequency and severity, refining the abilities of impacts models to handle the greater extremes in climate predicted by GCMs is also a fruitful area of research. For example, returning to the example of crop modeling, the CERES-Wheat model does not adequately take into account extreme daily climate events, such as the effect of high daily maximum temperatures (Rosenzweig, 1990; J. Ritchie, personal communication). It also cannot determine the effect of extremely high rainfall events (i.e., possible damage of too much precipitation). Such inadequacies could result in misconceptions about the effects of drought on various resource systems.

Further Analysis of Effect of Changing Variability on Drought

The paucity of research articles that directly deal with analysis of drought frequency in climate change experiments indicates that research efforts could be effectively directed toward more detailed analysis of climate modeling experiments already completed. This would entail more careful evaluation of the adequacy of soil moisture formulations in the models for serving as surrogates for drought and determining the adequacy of drought indices currently in use. The question still remains as to what the "best" indicator of drought in different climate models would be. It might be useful, for example, to establish thresholds of soil moisture decreases in modeling experiments which correspond to droughts of different magnitudes.

Furthermore, more analysis of variability changes on different time scales in GCM climate change experiments should be encouraged since only a handful of studies on this topic have been completed. With improvements in the climate models discussed above, it is likely that future efforts to quantify changes in drought severity and frequency will become more and more fruitful.

SUMMARY/CONCLUSION

In this chapter I have discussed the nature of drought as an extreme event, the ability of general circulation models to simulate drought, their ability to simulate aspects of climatic variability relevant to drought, and what we currently know from model results about how increased greenhouse gases in the atmosphere could alter the frequency and severity of drought. An example from research in crop-climate interactions is presented, which demonstrates the importance of changing climate variability as it affects crop failures resulting from moisture stress (the operational definition of drought in this context). In addition, I have attempted to cover what research could be undertaken in the future to improve our ability to understand and predict future occurrences of drought.

We do not know enough about the nature of drought as a climatic phenomenon, and climate models remain a long way from accurately reproducing drought episodes and properly modeling important aspects of interannual variability (of temperature and precipitation) that affect drought incidence and severity. This is one research area that has not received the kind of detailed attention from the modeling community that it deserves. It is an area that could greatly benefit from interdisciplinary research by climate modelers, observational climatologists, hydrologists, and climate impacts researchers.

ABOUT THE AUTHOR

Linda O. Mearns is a scientist in the Interdisciplinary Climate Systems Section of the National Center for Atmospheric Research, Boulder, Colorado. She holds a Ph.D. in geography/climatology from UCLA. She has done research and published in the areas of crop-climate interactions, climate change impacts on the biosphere, and statistics of climate variability and extreme climate events (in observations and climate models). She recently contributed to the *IPCC Climate Change 1992* report on the subject of climate variability in general circulation models.

REFERENCES

Barnett, T. P. 1985. Long-term changes in precipitation patterns. *In* M. C. MacCracken and F. M. Luther, eds. *Detecting the Climatic Effects of Increasing Carbon Dioxide*; pp. 149-62. U.S. Dept. of Energy, Carbon Dioxide Research Unit, Washington, D.C.

Boer, G. J.; K. Arpe; M. Blackburn; M. Deque; W. L. Gates; T. L. Hart; H. le Treut; E. Roeckner; D. A. Sheinin; I. Simmonds; R. N. B. Smith; T. Tokioka; R. T. Wetherald; and D. Williamson. 1992. Some results from an intercomparison of the climates simulated by 14 atmospheric general circulation models. *Journal of Geophysical Research* 97:12771-86.

Boer, G. J.; K. Arpe; M. Blackburn; M. Deque; W. L. Gates; T. L. Hart; H. le Treut; E. Roeckner; D. A. Sheinin; I. Simmonds; R. N. B. Smith; T. Tokioka; R. T. Wetherald; and D. Williamson. 1991. *An Intercomparison of the Climates Simulated by 14 Atmospheric General Circulation Models.* WMO/ICSU World Climate Research Program (WCRP-58). WMO, Geneva.

Boer, G. J.; and M. Lazare. 1988. Some results concerning the effect of horizontal resolution and gravity wave drag on simulated climate. *Journal of Climate* 1:789-806.

Boville, B. A. 1991. Sensitivity of simulated climate to model resolution. *Journal of Climate* 4:469-85.

Bradley, R. S.; H. F. Diaz; J. K. Eischeid; P. D. Jones; P. M. Kelly; and C. M. Goodess. 1987. Precipitation fluctuations over Northern Hemisphere land areas since the mid-19th century. *Science* 237:171-75.

Bretherton, F. P.; K. Bryan; and J. D. Woods. 1990. Time-dependent greenhouse gas induced climate change. *In* J. T. Houghton, G. T. Jenkins, and J. J. Ephraums, eds. *Climate Change: The IPCC Scientific Assessment*; pp. 177-93. Report prepared for the IPCC by Working Group I, University Press, Cambridge.

Brinkmann, W. A. R. 1983. Variability of temperature in Wisconsin. *Monthly Weather Review* 111:172-80.

Brklacich, M.; and B. Smit. 1992. Implications of changes in climatic averages and variability on food production opportunities in Ontario, Canada. *Climatic Change* 20:1-21.

Bryson, R. A. 1975. The lessons of climatic history. *Environmental Conservation* 2:163-70.

Cao, H.-X.; J. F. B. Mitchell; and J. R. Lavery. 1992. Simulated diurnal range and variation of surface temperature in a global climate model for present and doubled CO_2. *Journal of Climate* (in press).

Carter, T. R.; and M. L. Parry. 1986. Climatic changes and yield variability. *In* P. B. R. Hazell, ed. *Summary Proceedings of a Workshop on Cereal Yield Variability*; pp. 47-68. International Food Policy Research Institute, Washington, D.C.

Chico, T.; and W. D. Sellers. 1979. Interannual temperature variability in the United States since 1896. *Climatic Change* 2:139-47.

Cubasch, U. 1991. Preliminary assessment of the performance of a global coupled atmosphere-ocean model. *In* M. Schlesinger, ed. *Greenhouse-Gas-Induced Climatic Change: A Critical Appraisal of Simulations and Observations*; pp. 137-50. Elsevier, New York.

Diaz, H. F. 1983. Drought in the United States: Some aspects of major dry and wet periods in the contiguous United States, 1895-1981. *Journal of Climate and Applied Meteorology* 22:3-16.

Diaz, H. F.; and R. G. Quayle. 1980. The climate of the United States since 1895: Spatial and temporal changes. *Monthly Weather Review* 108:249-66.

Dickinson, R. E. 1989. Uncertainties of estimates of climatic change: A review. *Climatic Change* 15:5-13.

Dickinson, R. E.; and A. Henderson-Sellers. 1988. Modelling tropical deforestation: A study of GCM land-surface parameterizations. *Quarterly Journal of the Royal Meteorological Society* 114(B):439-62.

Druyan, L. M. 1988. Sea-surface temperature-Sahel drought teleconnections in GCM simulations. *In* S. Gregory, ed. *Recent Climatic Change*; pp. 154-65. Belhaven Press, London.

Druyan, L. M.; and S. Hastenrath. 1991. Modelling the differential impact of 1984 and 1950 sea-surface temperatures on Sahel rainfall. *International Journal of Climatology* 11:367-80.

Easterling, W. E.; N. J. Rosenberg; and M. S. McKenney. 1992. An introduction to the methodology, the region of study, and a historic analog of climate change. *Agricultural and Forest Meteorology* 59:3-16. Special issue: Methodology for Assessing Regional Agricultural Consequences of Climate Change.

Giorgi, F.; and L. O. Mearns. 1991. Approaches to the simulations of regional climate change: A review. *Reviews of Geophysics* 29:191-216.

Giorgi, F.; M. R. Marinucci; and G. Visconti. 1992a. A 2xCO_2 climate change scenario over Europe generated using a limited area model nested in a general circulation model. II. Climate change scenario. *Journal of Geophysical Research* 97:10011-28.

Giorgi, F.; C. S. Brodeur; and G. T. Bates. 1992b. Regional climate change scenarios over the United States produced with a nested regional climate model: Spatial and seasonal characteristics. Submitted to *Journal of Climate*.

Glantz, M.; R. Katz; and M. Krenz, eds. 1987. *The Societal Impacts Associated with the 1982-83 Worldwide Climate Anomalies*. UNEP, Nairobi; and ESIG/NCAR, Boulder, Colorado, USA.

Gordon, H. B.; and B. G. Hunt. 1991. Droughts, floods, and sea-surface temperature anomalies: A modeling approach. *International Journal of Climatology* 11:347-65.

Gordon, H. B.; P. Whetton; A. B. Pittock; A. Fowler; M. Haylock; and K. Henessy. 1992. Simulated changes in daily rainfall intensity due to the enhanced greenhouse effect: Implications for extreme rainfall events. Submitted manuscript.

Grotch, S. L. 1988. *Regional Intercomparisons of General Circulation Model Predictions and Historic Climate Data*. U.S. Department of Energy, Carbon Dioxide Research Division Report TR041. Washington, D.C.

Hansen, J.; I. Fung; A. Lacis; S. Lebedeff; D. Rind; R. Ruedy; G. Russell; and P. Stone. 1988. Global climate changes as forecast by the Goddard Institute for Space Studies three-dimensional model. *Journal of Geophysical Research* 93(D8):9341-64.

Houghton, J. T.; G. T. Jenkins; and J. J. Ephraums, eds. 1990. *Climate Change: The IPCC Scientific Assessment*. Report prepared for the IPCC by Working Group I. University Press, Cambridge.

Houghton, J. T.; B. A. Callander; and S. K. Varney, eds. 1992. *Climate Change 1992*. The Supplementary Report to the IPCC Scientific Assessment. University Press, Cambridge.

Hulme, M. 1991. An intercomparison of model and observed global precipitation climatologies. *Geophysical Research Letters* 18:1715-18.

Katz, R. W. 1992. Role of statistics in the evaluation of general circulation models. *Climate Research* 2:35-45.

Katz, R. W.; and B. G. Brown. 1992. Extreme events in a changing climate: Variability is more important than averages. *Climatic Change* 21:289-302.

Kellogg, W. W.; and Z. Zhao. 1988. Sensitivity of soil moisture to doubling of carbon dioxide in climate model experiments: Part I. North America. *Journal of Climate* 1:348-66.

Kiehl, J. T.; and D. L. Williamson. 1991. Dependence of cloud amount on horizontal resolution in the National Center for Atmospheric Research Community Climate Model. *Journal of Geophysical Research* 96(D):10955-80.

Kiladis, G. N.; and H. F. Diaz. 1989. Global climatic anomalies associated with extremes in the Southern Oscillation. *Journal of Climate* 2:1069-90.

Kitoh, A. 1992. Simulated interannual variations of the Indo-Australian monsoons. *Journal of the Meteorological Society of Japan* (in press).

Latif, M.; M. Flugel; and J.-S. Xu. 1991. *An Investigation of Short Range Climate Predictability in the Tropical Pacific*. Max Planck Institute of Meteorology, Report No. 52, Hamburg.

Lough, J. M.; T. M. L. Wigley; and J. P. Palutikof. 1983. Climate and climate impact scenarios for Europe in a warmer world. *Journal of Climate and Applied Meteorology* 22:1673-84.

Manabe, S.; M. J. Spelman; and R. J. Stouffer. 1992. Transient responses of a coupled ocean-atmosphere model to gradual changes of atmospheric CO_2. Part II: Season response. *Journal of Climate* 5:105-26.

Manabe, S.; and R. T. Wetherald. 1987. Large scale changes of soil wetness induced by an increase in atmospheric carbon dioxide. *Journal of Atmospheric Science* 44:1211-35.

Mearns, L. O. 1989. Climate Variability. *In* J. Smith and D. Tirpak, eds. *The Potential Effects of Global Climate Change on the United States*; pp. 29-54. U.S. EPA Report to Congress, No. 230-05-61-050. Washington, D.C.

Mearns, L. O.; R. W. Katz; and S. H. Schneider. 1984. Extreme high-temperature events: Changes in their probabilities with changes in mean temperature. *Journal of Climate and Applied Meteorology* 23:1601-13.

Mearns, L. O. 1992a. Changes in climate variability with climate change. *In* S. K. Majumdar, L. S. Kalkstein, B. Yarnel, E. W. Miller, and L. M. Rosenfeld, eds. *Global Climate Change: Implications, Challenges, and Mitigation Measures*. Pennsylvania Academy of Sciences (in press).

Mearns, L. O. 1992b. Changing climate variability with greenhouse warming and its possible impacts. Paper presented at the First Nordic Inter-disciplinary Conference on the Greenhouse Effect, Copenhagen, 16-18 September 1991 (Proceedings in press).

Mearns, L. O.; S. H. Schneider; S. L. Thompson; and L. R. McDaniel. 1990. Analysis of climate variability in general circulation models: Comparison with observations and changes in variability in 2xCO_2 experiments. *Journal of Geophysical Research* 95:20469-90.

Mearns, L. O.; C. Rosenzweig; and R. Goldberg. 1992a. The effect of changes in interannual climatic variability on CERES-Wheat yields: Sensitivity and 2xCO_2 GCM studies. *Journal of Agricultural and Forest Meteorology* (in press).

Mearns, L. O.; C. Rosenzweig; and R. Goldberg. 1992b. *Sensitivity Analysis of the CERES-Wheat Model to Changes in Interannual Climate Variability.* Report to the U.S. EPA, Office of Planning and Policy Evaluation, Global Climate Change Division (in press).

Meehl, G. A. 1990. Seasonal cycle forcing of El Niño-Southern Oscillation in a global coupled ocean-atmosphere GCM. *Journal of Climate* 3:72-98.

Meehl, G. A.; G. W. Brantstator; and W. M. Washington. 1992. Tropical Pacific interannual variability and CO_2 climate change. Submitted to *Journal of Climate.*

Mitchell, J. F. B.; C. A. Wilson; and W. M. Cunnington. 1987. On CO_2 climate sensitivity and model dependence of results. *Quarterly Journal of the Royal Meteorological Society* 113:293-322.

Mitchell, J. F. B.; C. A. Senior; and W. J. Ingram. 1989. CO_2 and climate: A missing feedback. *Nature* 341:132-34.

Mitchell, J. F. B.; S. Manabe; V. Meleshko; and T. Tokioka. 1990. Equilibrium climate change—and its implications for the future. *In* M. Schlesinger, ed. *Greenhouse-Gas-Induced Climatic Change: A Critical Appraisal of Simulations and Observations*; pp. 17-34. Elsevier, New York.

Oglesby, R. J. 1991. Springtime soil moisture, natural climatic variability, and North American drought as simulated by the NCAR Community Climate Model 1. *Journal of Climate* 4:890-97.

Oglesby, R. J.; and D. J. Erickson. 1989. Soil moisture and the persistence of North American drought. *Journal of Climate* 2:1362-80.

Oglesby, R. J.; and B. Saltzman. 1990. Sensitivity of the equilibrium surface temperature of a GCM to systematic changes in atmospheric carbon dioxide. *Geophysical Research Letters* 17:1089-92.

Palmer, T. N.; C. Brankovic; P. Viterbo; and M. J. Miller. 1992. Modelling interannual variations of summer monsoons. *Journal of Climate* 5:399-417.

Panturat, S.; and A. Eddy. 1989. Some impacts on rice yields from changes in the variance of precipitation. *Climate Bulletin* 24:16-27.

Parry, M. L.; and T. R. Carter. 1985. The effect of climatic variations on agricultural risk. *Climatic Change* 7:95-110.

Philander, S. G. H.; R. C. Pacanowski; N-C. Lau; and M. J. Nath. 1992. A simulation of the Southern Oscillation with a global atmospheric GCM coupled to a high-resolution tropical Pacific ocean GCM. *Journal of Climate* (in press).

Pitcher, E. J.; M. L. Blackmon; M. L. Gates; G. T. Bates; and S. Munoz. 1988. The effect of north Pacific sea surface temperature anomalies on the January climate of a general circulation model. *Journal of Atmospheric Science* 45:173-88.

Pittock, A. B.; A. M. Fowler; and P. H. Whetton. 1991. Probable changes in rainfall regimes due to the enhanced greenhouse effect. *Proceedings of the International Hydrology and Water Resources Symposium*, Perth, Australia, October 1991.

Rind, D. 1986. The dynamics of warm and cold climates. *Journal of Atmospheric Science* 43:3-24.

Rind, D. 1991. Climate variability and climate change. *In* M. Schlesinger, ed. *Greenhouse-Gas-Induced Climatic Change: A Critical Appraisal of Simulations and Observations*; pp. 69-78. Elsevier, New York.

Rind, D.; R. Goldberg; J. Hansen; C. Rosenzweig; and R. Ruedy. 1990. Potential evapotranspiration and the likelihood of future drought. *Journal of Geophysical Research* 95(D7):9983-10004.

Rind, D.; R. Goldberg; and R. Ruedy. 1989. Change in climate variability in the 21st century. *Climatic Change* 14:5-38.

Rosenzweig, C. 1990. Crop response to climate change in the Southern Great Plains: A simulation study. *The Professional Geographer* 42:20-39.

Schuurmans, C. J. E.; and A. J. Coops. 1984. Seasonal mean temperatures in Europe and their interannual variability. *Monthly Weather Review* 112:1218-25.

Sellers, P. J.; Y. Mintz; Y. C. Sud; and A. Dalcher. 1986. A simple biosphere model (SiB) for use within general circulation models. *Atmospheric Science* 43:505-31.

Skeeter, B. R.; K. E. Dewey; and J. A. Harrington. 1988. A climatology of intra-monthly temperature variability in the conterminous United States 1951-1988. *Physical Geography* 9:99-119.

Slingo, A.; and J. M. Slingo. 1991. Response of the National Center for Atmospheric Research Community Climate Model to improvements in the representation of clouds. *Journal of Geophysical Research* 96(D):15341-57.

Smith, J. B.; and D. A. Tirpak, eds. 1989. *Report to Congress on the Potential Effects of Global Climate Change on the U.S.* EPA-220-05-89-050, EPA, Washington, D.C.

Stouffer, R. J.; S. Manabe; and K. Bryan. 1989. Interhemispheric asymmetry in climate response to a gradual increase of atmospheric CO_2. *Nature* 342:660-62.

Trenberth, K. E.; G. W. Brantstator; and P. Arkin. 1988. Origins of the 1988 North American drought. *Science* 242:1640-1645.

Tyson, P. D. 1977. The enigma of changing world climates. *South African Geographical Journal* 59:77-116.

van Loon, H.; and J. Williams. 1978. The association between mean temperature and interannual variability. *Monthly Weather Review* 106:1012-17.

Verseghy, D. A. 1991. Class—A Canadian land surface scheme for GCMs. I. Soil model. *International Journal of Climatology* 11:111-34.

Washington, M.; and G. A. Meehl. 1989. Climate sensitivity due to increased CO_2: Experiments with a coupled atmosphere and ocean general circulation model. *Climate Dynamics* 4:1-38.

Whetton, P. Unpublished maps of daily and interannual temperature variability change for the region of Australia. Analyses of the CSIRO climate model doubled CO_2 experiments.

Wigley, T. M. L. 1985. Impact of extreme events. *Nature* 316:106-7.

Wigley, T. M. L. 1988. The effect of changing climate on the frequency of absolute extreme events. *Climate Monitor* 17:44-55.

Wilks, D. S. 1992. Adapting stochastic weather generation algorithms for climate change studies. *Climatic Change* 22:67-84.

Wilson, C. A.; and J. F. B. Mitchell. 1987. Simulated climate and CO_2-induced climate change over Western Europe. *Climatic Change* 10:11-42.

Zhao, Z.; and W. W. Kellogg. 1988. Sensitivity of soil moisture to doubling of carbon dioxide. Part II. The Asian monsoon region. *Journal of Climate* 1:367-88.

Part Two

Case Studies

Chapter 8
Drought Risk and Water Management in Israel: Planning for the Future

Hendrik J. Bruins

INTRODUCTION

The country of Israel is prone to significant variations in the annual amount of precipitation because of its geographical position at the northern edge of the largest desert belt on earth. The Sahara and Arabian-Iranian deserts make up a continuous arid zone, which stretches from the Atlantic coast of West Africa into central Asia over a distance of some 8,000 km. The southern half of Israel, the Negev, forms part of this planetary desert belt. The transition from hyperarid to subhumid occurs in Israel over a distance of merely 100-150 km. It is not surprising, therefore, to find drought as a recurring phenomenon in the region.

Certain aspects of drought in Israel have been taken into consideration in regional agricultural planning. Farmers who grow wheat in the northern Negev are compensated for drought, according to certain regulations. Economic aspects of this form of drought compensation in wheat production were studied by Bigman et al. (1987) and Becker (1987), who concluded that it could be useful with the imposition of appropriate restrictions.

However, there appears to be very little evidence of concrete and comprehensive planning for drought on a national scale. Virtually all the available freshwater resources in Israel are already used to the maximum and beyond, which has caused an alarming drop in potable groundwater reserves. If a severe meteorological drought would strike the region in the near future, the impact could be disastrous, unless new water resource management and proactive drought planning is developed and implemented as soon as possible.

THE PHENOMENON OF DROUGHT

Although drought is not a sudden violent event like an earthquake or a hurricane, its impact can, nevertheless, be devastating. One of the reasons for drought to be perceived as nondramatic, in comparison to other natural disasters, is its passive nature. Drought occurs because the rains fail to come, not because something happens. Since drought usually develops in a gradual way, it has been described by Gillette (1950) as a "creeping phenomenon." Prediction of the beginning as well as the end of a drought is a difficult task (Wilhite and Glantz, 1985). The impact of drought on society evolves gradually and occurs as direct and indirect effects (Parry and Carter, 1987; Riebsame, 1988).

The phenomenon of drought may be analyzed in terms of meteorological, agricultural, hydrological, socioeconomic, and organizational aspects (Wilhite and Glantz, 1985). The possible influence of man on the occurrence of drought was expressed by Hoyt in 1942 (cited in Wilhite and Glantz, 1985): Drought may result if "in the economic development of a region man creates a demand for more water than is normally available." This aspect of manmade drought clearly fits the current water problems in Israel.

The impact of drought must be seen as dynamic, basically as the result of interactions between supply and demand. The relationship between the supply of water, which can be expressed in physical as well as organizational terms, and the demand for water (socioeconomic criteria) is clearly not static and may vary considerably with time (Wilhite and Glantz, 1985). From a comparative study about drought policy in the United States and Australia, carried out by Wilhite (1986), it became clear that governments often respond to drought through crisis management rather than preplanned programs (i.e., risk management). Hastily prepared assessment and response procedures may lead to ineffective, poorly coordinated, and untimely response (Wilhite and Wood, 1985; Wilhite et al., 1986; Wilhite and Easterling, 1987).

CLIMATE AND WATER RESOURCES OF ISRAEL

Climate

The climate of Israel is characterized by a hot dry summer and a cool rainy winter. Average annual precipitation ranges from about 900 mm in the north to only 25 mm in the south. About half the country is made up of deserts. A meaningful numerical classification of the climate in Israel (Table 1) can be presented accurately and conveniently on the basis of the P/ETP index (P = average annual precipitation; ETP = average annual potential evapotranspiration). This system of climatic classification for arid regions is also used by UNESCO (1979).

The climate in Israel ranges, according to this classification system, from hyperarid to subhumid, although a few wet mountain spots in the north are classified as humid. The P/ETP index varies correspondingly from 0.01 near Eilat in the hyperarid southernmost part of the Negev to about 0.80 in the wettest parts of the Galilee in the north of the country. The prevailing aridity in large parts of Israel emphasizes the necessity of careful management of water resources.

CLIMATE	P/ETP Index	Geographic Region
Hyperarid	< 0.03	Southern and Eastern Negev, Arava Valley, Dead Sea
Arid	0.03-0.20	Central and Northern Negev, Judean desert, Lower Jordan Valley
Semiarid	0.20-0.50	Southern central part of the country, Kinneret and Bet Shean Valley
Subhumid	0.50-0.75	Approximately north of the line from Tel Aviv to Jerusalem
Humid	> 0.75	Highest mountain areas in the north of the country

Table 1. Climatic aridity in Israel.

Water Resources

The main water resources of Israel are concentrated in three principal catchment basins: Lake Kinneret (Figure 1:1), the Coastal Aquifer (Figure 1:3), and the Yarqon-Tanninim Aquifer (Figure 1:2). About 80% of all the water used in the country is derived from these three catchment basins. Other important water resources are formed by the aquifers of Western Galilee (Figure 1:4,5), the Carmel region (Figure 1:6), the Eastern Galilee (Figure 1:1), the Golan (Figure 1:1), the Yizreel Valley (Figure 1:7[1]) and Bet Shean Valley (Figure 1:7[1]), and the aquifers in the Negev (Figure 1:8) and the Arava Valley (Figure 1:9) (Gilead and Bachmat, 1973; Grinwald and Bibas, 1989).

The distribution of surface and groundwater in Israel is largely a reflection of the climatic situation: most of the water resources are concentrated in the northern part of the country. It became clear within the first decade after the establishment of the state that the various regional water projects would not be adequate to meet the water needs of the country in the years ahead. As a result, the National Water Carrier (Figure 2) was built, the largest single development project undertaken in Israel. Construction of the system began in 1953 and was completed in 1964. Its principal function is to pump water from Lake Kinneret and transport it south. An average amount of 450 million cubic meters (mcm) is pumped from the lake annually. However, the National Water Carrier system integrates most of the regional water resources in Israel, and it also has an important seasonal and interannual regulation function. In the rainy winter season, it is used to transport surplus water to underground reservoirs for storage in the aquifers as buffer stocks for the dry summer season and to store excess water from wet years for use during drought years (Ministry of Agriculture, 1973). However, in reality, too little water has so far been returned to make up for the overexploitation of the aquifers in the last few decades.

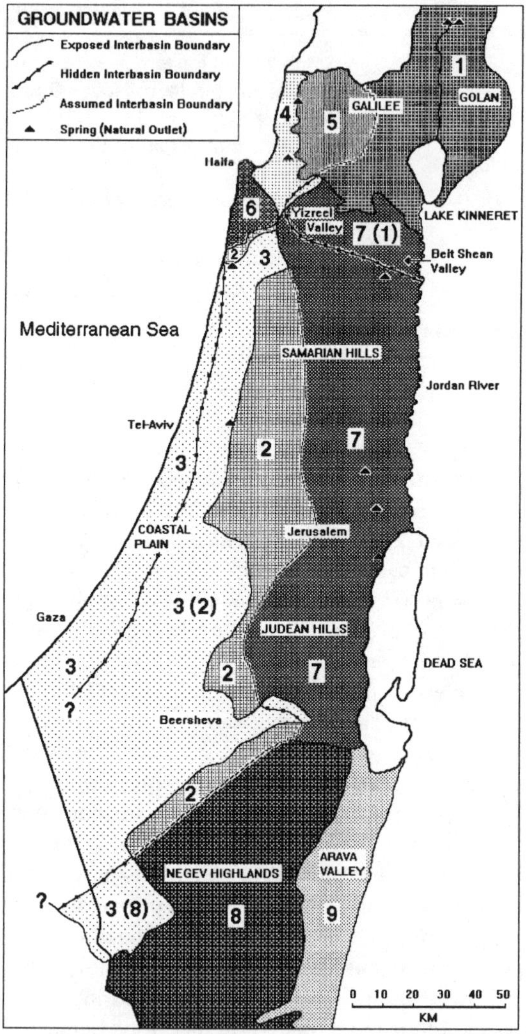

Figure 1. Map of groundwater basins (modified after Gilead and Bachmat, 1973): 1. Basin of Lake Kinneret and the northern Jordan Valley; 2. Western basin of the Samarian and Judean hills (Yarqon-Tanninim Aquifer), as well as the northwestern fringe basin of the Negev hills; 3. Coastal plain basin; 4. Coastal basin of Western Galilee and Haifa bay; 5. Western Galilee hill basin; 6. Carmel basin; 7. Eastern basin of the Samarian and Judean hills; 8. Negev Highlands basin; 9. Basin of the Arava Valley.

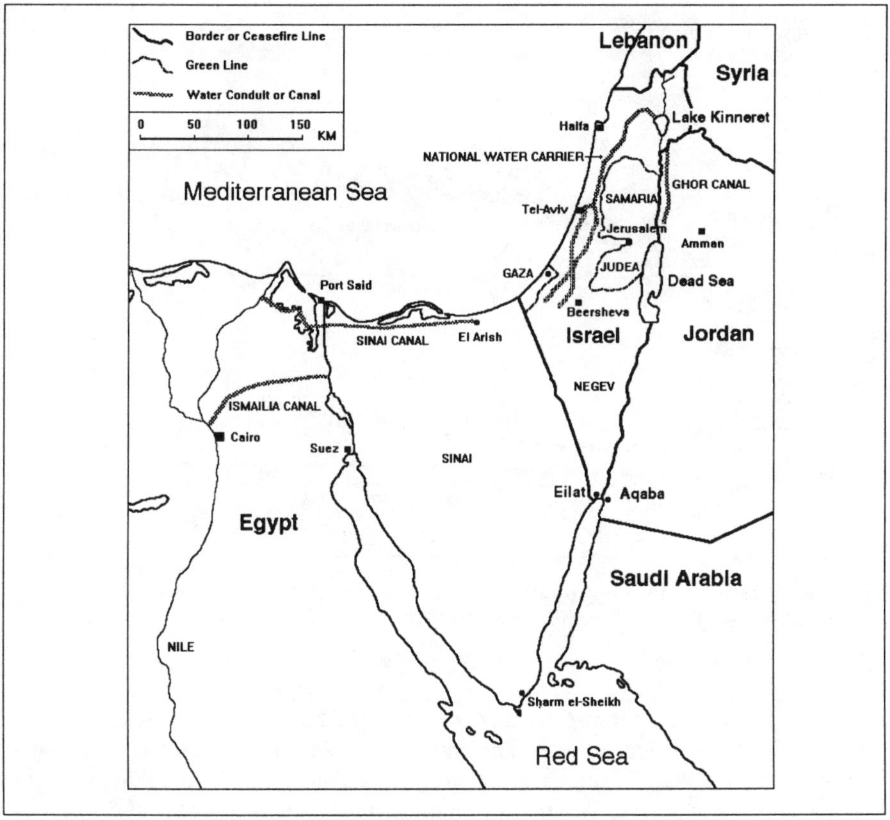

Figure 2. Regional map with existing conduits or canals for water transportation in Israel, Egypt and Jordan (partly based on Kally, 1989).

Total water consumption in Israel in the last 10 years has ranged from a high of 2,024 mcm in 1985 to a low of 1,442 mcm in 1991 (Bibas, personal communication, 1992; Figure 3). The latter drop, in 1991, was caused by the rather sudden public realization of the alarming state of Israel's dwindling water resources, following a report by the State Comptroller (1990). The authorities reacted by enforced cutbacks in water supply, particularly in the agricultural sector.

Lake Kinneret. The only large natural storage of fresh surface water in the country is formed by Lake Kinneret, which is also known as Lake Galilee. Water is supplied to the lake by the Jordan River and its tributaries, and to a lesser extent by wadis and springs

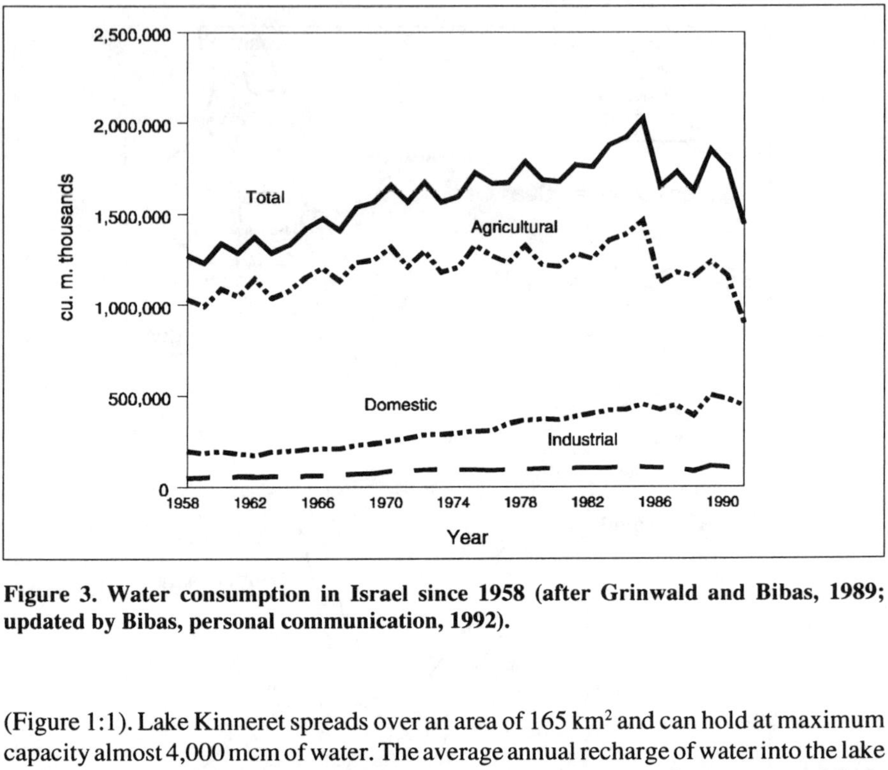

Figure 3. Water consumption in Israel since 1958 (after Grinwald and Bibas, 1989; updated by Bibas, personal communication, 1992).

(Figure 1:1). Lake Kinneret spreads over an area of 165 km^2 and can hold at maximum capacity almost 4,000 mcm of water. The average annual recharge of water into the lake amounts to about 600 to 530 mcm. The average annual amount of water pumped from the lake into the National Water Carrier system is about 450 mcm (Ministry of Agriculture, 1973; Grinwald and Bibas, 1989).

The salt content of Lake Kinneret was about 400 mg/l (or ppm), which is above the accepted national standard and World Health Organization guideline (WHO, 1988) of 250 mg/l. However, the authorities were able to divert saline water from springs discharging into the lake. Thus the salinity of Lake Kinneret was brought down to acceptable levels, between 205 and 230 mg/l, which is a remarkable achievement (Ministry of Agriculture, 1973; Grinwald and Bibas, 1989).

Lake Kinneret is situated in the rift valley, and its surface and the surrounding shore lie more than 200 m below mean ocean level. At maximum capacity, the lake level stands at -208.9 m; the minimum acceptable level has been determined at -213 m, the so-called red line. Below this red line, water quality in the lake, particularly salinity, is expected to deteriorate to unacceptable levels. Each meter in the water level of Lake Kinneret represents about 170 mcm of water. The overflow into the Jordan River at the southern end of the lake is controlled by a sluice gate, through which the lake level can be maintained at the desired levels. However, during the extremely wet winter of 1991-92, the maximal outflow proved almost too small to prevent the lake from flooding the surrounding settlements. In the same winter, the lake level rose dramatically from its lowest level in recent history, below the red line of -213 m, to its maximum level of -208.9 m.

Coastal Aquifer. The Coastal Aquifer is situated below the coastal plain along the Mediterranean Sea (Figure 1:3). Its length from north to south is about 140 km and its average width is only some 15 km. The total area of the Coastal Aquifer comprises approximately 1,800 km^2. Its rocks are composed of Late Tertiary to Quaternary sand and sandstone with intervening layers of clay and fossil soils, which bring about divisions in a number of subaquifers (Issar, 1968). The sandy deposits attain a thickness of about 120 m near the coast, gradually becoming thinner toward the east. The basis of the aquifer is formed by a thick deposit of Tertiary clays.

Use of the Coastal Aquifer started at the beginning of this century, and until the 1950s it supplied most of the water used in the country. The amount of wells gradually increased and so did the amount of water pumped from the aquifer: about 250 mcm in 1948 and a record 493 mcm in 1958. With the opening of the National Water Carrier in 1964, less water was initially withdrawn from the Coastal Aquifer. However, pumping later increased again, reaching 470 mcm in the year 1984-85 (Grinwald and Bibas, 1989). The continuous overpumping of the coastal aquifer has resulted in a drop in the water table, increased penetration of seawater, and a general reduction in water quality (Melloul and Bibas, 1990).

Yarqon-Tanninim Aquifer. The Yarqon-Tanninim aquifer is composed of hard calcareous rocks of Cenomanian-Turonian age, forming part of the geology of the central hills (Figure 1:2). Its main reserve is situated in the Yarqon-Tanninim basin east of the coastal plain. The annual replenishment in the basin has been estimated at about 360 mcm for the area between Pardes Hanna and Lod (Kally, 1973). The aquifer extends south to Beersheva and is of a karstic nature with a high conductivity and swift flows (Grinwald and Bibas, 1989).

Use of the Yarqon-Tanninim aquifer has increased rapidly since the 1950s and has reached annual withdrawal levels of more than 400 mcm in recent years. These figures are higher than the long-term safe yield estimates of 340 mcm per year. Water levels in the aquifer have dropped by about 7 m as a result, and the water quality has deteriorated. Every meter in the level of the aquifer represents an estimated amount of 100 mcm water. Further lowering of the water level in the aquifer may introduce salinity problems in this largest groundwater reservoir of the country.

The Yarqon-Tanninim aquifer constitutes the principal source of fresh water in Israel. It functions as a seasonal and interannual buffer, in conjunction with Lake Kinneret and the National Water Carrier system, to ensure water supply in the country.

The amount of storage in the aquifer between the uppermost level, in which the water escapes through natural springs, and the lowest level or red line amounts to about 900 mcm. Preservation of its water quality is of great importance.

WATER MANAGEMENT AND THE MANMADE DROUGHT

Legal and Organizational Aspects

Water resources in Israel are centrally governed by the state, based on the Water Law enacted in August 1959. The first clause of the law states: "The water resources in the State are public property, under the control of the State, and intended for the needs of its residents and the development of the country." The minister of agriculture is responsible for and in charge of implementation of the water legislation. A water commissioner is appointed under the minister of agriculture in order to manage the water affairs of the state through a water commission. The latter body consists of five main departments: (1) Allocation and Licensing, (2) Drainage, (3) Hydrological Service, (4) Efficient Use of Water, and (5) Head Office (Ministry of Agriculture, 1973; Grinwald and Bibas, 1989).

Mekorot Water Company Limited operates the water supply of the country and also has constructed its infrastructure. The building of the National Water Carrier has been one of its major achievements. The company, in which the government is one of the shareholders, was founded in 1937. TAHAL is the National Water Planning Company, a public corporate body, established in 1952. It functions as the government planning agency.

Public participation in matters related to water is organized through a water board, which has a maximum number of 39 representatives. The Water Board has no power of decision. Its main function is to give advice and recommendations to the minister of agriculture (Ministry of Agriculture, 1973).

The Manmade Drought

With the publication of a report by the state comptroller (1990) in the first days of January 1991, the general public was suddenly alerted, through the considerable degree of media attention, to the grave situation of Israel's water resources. The *Jerusalem Post* (January 3, 1991) dedicated its entire front page to the subject under the banner headline: "Ben-Porat: Water catastrophe looming." In her 62-page report, state comptroller Miriam Ben-Porat concluded that the country is on the verge of a catastrophic water shortage after 25 years of irresponsible management, which has resulted in a very serious overdraft from the aquifers of 1,600 mcm of water. This deficit is equivalent to the water requirement of the country for one year. The vital water reserves in Israel have been depleted to dangerously low levels, and the water quality has been seriously damaged.

The report warns of the very real danger that Israel may not be able to supply enough minimal quality water to satisfy demand in the near future. Some of the groundwater reservoirs in the country may have been damaged irreparably. The findings of the report have been summarized succinctly by Hellman and Rudge (1991) in the *Jerusalem Post*, quoted hereby at length:

> Ben-Porat blames the catastrophe on a system that permits interested parties, such as the agricultural lobby, through its control of the powerful Knesset Water Committee, to make major decisions regarding the pricing, subsidizing, and allocation of water. The comptroller therefore recommends that water management be removed from the Agriculture Ministry and put under the control of an objective state authority. Such a body would be concerned with the country's overall needs, and would not be influenced by vested interest groups... The report stresses that previous agriculture ministers and water commissoners had deliberately ignored dire warnings of an impending water catastrophe, and that their decision to ignore recommendations voiced repeatedly over the past 25 years was 'erroneous and highly dangerous to the state's water resources.'

Ben-Porat's findings also hint at possible violations of laws regulating water management, in that officers ignored statistics presented to them. They persistently overestimated the amount of water available by counting sewage water as a potential reserve even though many of the planned sewage treatment facilities have not yet been established. Mekorot, the national water carrier, and the State Water Commission's own hydrological service are not blamed by the report, which notes that these bodies were among those that warned of impending shortages.

A particular concern expressed in the report refers to the low price charged to agricultural concerns for water consumption. The report blames the policy by the Ministry of Agriculture of subsidizing agricultural water consumption for much of the current crisis. Water subsidies were passed on to foreign consumers through cheap agricultural export prices, leading to what the report terms "essentially, the export of water at a loss" (Hellman and Rudge, 1991).

Almost 20 years ago, Kally (1973) wrote about the danger of mining the one-time reserve in the Yarqon-Tanninim aquifer: "The annual replenishment of this basin is on the average about 360 mcm (million cubic metres), but it contains billions of cubic metres of water that has accumulated in the past and can now be drawn by available technical means. So far, no steps have been taken to reduce agricultural water consumption... If consumption meanwhile continues growing at the present rate, and does so up to the end of the present decade [*until 1980*], the national reserve will be depleted by an amount of the order of magnitude of 1,000 mcm (assuming average climatic conditions)." The forecast and warning of Kally proved to be correct, with the added difference that by 1991 the deficit had risen to 1,600 mcm, while the reserves in the aquifer have dwindled below the red line.

METEOROLOGICAL DROUGHT IN ISRAEL

Three main types of meteorological drought may be recognized in Israel: (1) seasonal drought; (2) frequently occurring annual to multi-annual drought of light to medium severity; and (3) low-frequency multi-annual drought of great severity, known from biblical and other historical literary sources.

Seasonal Drought

The annual climate is characterized by a distinct seasonality. In the winter the country is effected by the middle-latitude belt of the so-called westerlies, which bring cyclonic rains to the Levant. However, in the summer the entire country comes under the influence of subsiding air of the so-called Hadley cell, causing stable blue skies. The summer is usually completely dry for about five months. This seasonal drought recurs every year with the regularity of an annual meteorological clock.

Annual to Multi-Annual Drought of Light to Medium Severity

Israel lies near the southern limit of cyclonic winter rains. The number and trajectories of these depressions often determine the amount of rainfall. Large variations in the amount of interannual precipitation occur over Israel (Katsnelson, 1964), resulting in the regular occurrence of temporal meteorological drought.

Instrumented rainfall records are available for Jerusalem since 1846. Zangvil (1979) studied a series of 108 years of rainfall data of Jerusalem for the period 1846-47 to 1953-54. He determined by time spectrum analysis and filtering techniques that the most prominent feature is a peak at a period of 3.0 to 3.3 years. Secondary peaks occur near 2.15 and 6.0 years. Similarities were noted with certain rainfall variations and patterns over California, which seem to be almost in phase in both regions, apparently affected by the same global climatic controls. It was suggested by Zangvil (1979) that the strong peak of 3.0 to 3.3 years and the weak peak of 6.0 years may be associated with the Southern Oscillation.

An analysis of drought years for three experimental farms in the Negev showed the occurrence of drought years with a periodicity of about 3.2 to 3.5 years (Table 2) (Bruins, 1986; Bruins et al., 1986 and 1987). These results are similar to the main periodicity

Rainy Season	Avdat—Hillside Conduit System			Shivta—Hillside Conduit System			Mashash—Liman System		
	Rain (mm)	Runoff (m³)	Drought Yr.	Rain (mm)	Runoff (m³)	Drought Yr.	Rain (mm)	Runoff (m³/ha)	Drought Yr.
1960-61	70.0	1,932		102.5	594				
1961-62	64.7	4,471		50.6	280	*			
1962-63	29.5	792	*	24.8	44	*			
1963-64	183.3	51,875		145.3	6,293				
1964-65	144.2	6,115		160.8	9,444				
1965-66	91.5	6,905		67.1	1,297				
1966-67	80.7	1,827		94.4	2,285				
1967-68	86.9			75.4					
1968-69	70.3			73.2					
1969-70	58.4		*	43.6		*			
1970-71	82.0			115.2			142.0	7,000	
1971-72	162.6			172.4			207.0	10,000	
1972-73	53.5		*	41.0		*	55.0	0	*
1973-74	132.6			105.8			180.5	4,000	
1974-75	84.9			77.7			93.0	3,000	
1975-76	77.1			62.6		*	79.5	0	*
1976-77	64.1		*				101.0	2,000	
1977-78	68.0						75.0	1,500	
1978-79	56.0		*				90.7	5,000	
1979-80	100.6						166.0	16,500	
1980-81	100.8						120.7	10,000	
1981-82	68.8		*				75.5	0	*
1982-83	124.2						170.0	3,500	
1983-84	29.3		*				84.5	0	*
Average	86.8	10,560		88.3	2,891		117.2	4,464	
Drought Year Frequency (once every . . . years)			3.4			3.2			3.5

Table 2. Rainwater-harvesting drought years in the Negev (after Bruins, 1986). The three experimental runoff farms were established by Evenari, Shanan, and Tadmor (1982). The *hillside conduit system* at the Avdat and Shivta farms is based on the collection of runoff rainwater from the surrounding landscape through a series of simple conduits built from earth and stones on the hillsides. The *liman system* at the Mashash farm collects runoff from dry river beds (wadis) entering into a broad valley plain. A liman is an agricultural field surrounded by a check-dam or bund. The name liman is derived from the Greek word *limnos* (= lake). A liman filled with runoff floodwater indeed appears as a little lake for a few days, until the water has infiltrated into the soil (Bruins, 1986; Bruins et al., 1986).

of 3.0 to 3.3 years found by Zangvil (1979) for the rainfall variations in Jerusalem. These rather frequently occurring droughts may be classified in the range of light to medium severity. Sometimes drought years are of multi-annual (rather than annual) duration, like the prolonged drought of 1959-60 and 1960-61, which was undoubtedly of medium severity.

Low-Frequency Multi-Annual Drought of Great Severity

The instrumental meteorological record is obviously limited to the recent past. In order to enlarge the time range to study climatic variations and drought, one has to look for proxy paleoclimatic indicators. The Dead Sea is undoubtedly the best paleoclimatic indicator in the southern Levant (Klein, 1961, 1981, 1982, and 1986; Neev and Emery, 1967; Begin et al., 1985; Frumkin et al., 1991). Bruins (1992) related climatic changes of the past 5,500 years to human history, based on three different time scales, as a basis for climate impact assessment.

The lake level changes of the Dead Sea have been reconstructed by Klein for the last 1,000 years in remarkable, almost annual, detail. Such a degree of resolution cannot usually be attained in paleoclimatic studies. Only medium to long-term climatic changes can be reconstructed on the basis of available data. The resolution of proxy paleoclimatic indicators is usually limited and below the required degree of detail to pinpoint individual drought years or periods of severe drought. Tree ring analysis might be able to contribute in principle, but data interpretation is still ambiguous in terms of cause and effect, so that the drought factor cannot be isolated easily from other climatic factors and soil moisture storage. Moreover, the time range covered by tree rings is rather limited for the Near East. Ancient literary data, on the other hand, can provide information about severe droughts in the past. In this respect, Israel has a unique data set in the Bible and later Judaic writings.

In the book of Genesis, several famines are mentioned during the time of the Patriarchs, usually associated with the Middle Bronze Age in archaeological cultural terms. With respect to absolute dating, opinions differ (Bruins and Mook, 1989), but the time of the Patriarchs has sometimes been associated with the early part of the second millennium B.C. The term *drought* is not mentioned as such in the biblical narrative, but it may be inferred that drought was the predominant cause for the occurrence of famine in those times. The Middle Bronze Age seems to have coincided with one of the driest climatic periods within historical times (Bruins, 1992), as indicated by the low level of the Dead Sea (Frumkin et al., 1991). However, one has to distinguish between an overall dry climatic period of several hundred years duration, in which common agricultural land use and pastoralism was still possible, and specific drought years within this period. The latter drought years, which caused famine, cannot be determined from proxy paleoclimatic indicators, but only from literary data.

"And there was a famine in the land; and Abraham went down into Egypt to sojourn there; for the famine was grievous in the land" (Genesis 12:10). Later, another famine occurred in the life of Isaac: "And there was a famine in the land, beside the first famine that was in the days of Abraham" (Genesis 26:1). A very severe famine of seven years' duration took place in the days of Jacob, which not only affected the land of Canaan but also the hydraulic Nile civilization of Egypt. Water supply and agriculture in hyperarid Egypt are based on extraterritorial water transported into the country by the Nile: hence the term *hydraulic civilization*. It seems again very likely that this large regional famine was caused by drought over the Levant as well as the catchment area of the Nile in central and East Africa. "For the famine has been in the land these two years; and there are yet five years in which there will be neither ploughing nor harvest" (Genesis 45:6). The latter verse clearly indicates that drought was the cause of the famine in the days of Jacob and Joseph. Ploughing in traditional agriculture could only be carried out in the land of Canaan after the first autumn or early winter rains had moistened the soil sufficiently to make it soft and suitable for ploughing and sowing. In traditional Egyptian agriculture, the farmers would plough and cultivate their lands only after the Nile floods had come and the water had subsided. The above verse, therefore, is quite convincing evidence of a meteorological drought in the land of Canaan and a hydrological drought in Egypt. "Now there was no food in all the land; for the famine was very severe, so that the land of Egypt and the land of Canaan languished by reason of the famine" (Genesis 47:13).

In a later period before the first kings, a famine is mentioned in the Bible: "In the days when the judges ruled there was a famine in the land, and a certain man of Bethlehem in Judah went to sojourn in the country of Moab, he and his wife and his two sons" (Ruth 1:1). In the time of King David, usually attributed to the eleventh to tenth century B.C., a famine occurred with a duration of three years: "Now there was a famine in the days of David for three years, year after year; and David sought the face of the LORD"(II Samuel 21:1).

About a century later, a severe drought hit the land in the days of King Ahab, whose reign is dated to the period of about 871-851 B.C. "And Elijah the Tishbite, who was of the inhabitants of Gilead, said unto Ahab, As the LORD God of Israel liveth, before whom I stand, there shall not be dew nor rain these years, but according to my word" (I Kings 17:1)."And it came to pass after many days, that the word of the Lord came to Elijah in the third year, saying, Go, show thyself unto Ahab; and I will send rain upon the earth... And there was a sore famine in Samaria" (I Kings 18:1-2). The combined situation of drought and famine are specifically mentioned in the latter verses. The duration of this severe drought is also spelled out specifically in a later reference written in the first century A.D., in the letter of James (5:17-18): "Elijah was a man of like nature with ourselves and he prayed fervently that it might not rain, and for three years and six months it did not rain on the earth. Then he prayed again and the heaven gave rain, and the earth brought forth its fruit." A considerable number of droughts in the country are recorded in the period 200-400 A.D. (Sperber, 1978), as reflected in the writings of contemporary Rabbinic sources.

PLANNING FOR SEVERE METEOROLOGICAL DROUGHT IN THE NEAR FUTURE

A Future Scenario Based on Ancient Data

If a severe meteorological drought would hit Israel in the near future, on top of the current manmade drought, the impact could indeed be catastrophic. The complexity of geophysical systems and the consequent random elements in all risk analyses preclude accurate drought prediction by deterministic methods. However, in a wise approach, the possible occurrence of severe droughts should be taken into consideration in the management of Israel's water resources. On the basis of paleoclimatic literary data, reviewed above, a future drought could be as severe as in the past. No climatic forecasting model would be able to predict when it might happen, but it could occur in the near future. A severe drought of three to four years duration, as occurred in the days of King David and later in the days of the prophet Elijah, is adopted as a future scenario for drought-planning purposes.

Groundwater Reserves for a Severe Drought

An important element of any drought planning in Israel must be the formation of a significant groundwater reserve for use in times of severe drought. Unfortunately, bad water resource management during the past 25 years has resulted in a deficit of 1.6 billion cubic meters of water. The reversal of the present deficit and the formation of a significant groundwater reserve can only be achieved through a drastic reduction in groundwater extraction. The amount of water required to sustain the population through a severe drought of about 3 to 4 years' duration is about 7,000 mcm, taking the maximum past annual water consumption of 2,000 mcm in 1985 as a basis for the calculation. However, this amount seems to exceed the potential for groundwater storage, if the present deficit of 1,600 mcm is taken as a starting point. It is not clear from available data how much water could be stored as groundwater, but it would seem to be not much more than the current deficit.

Some 20 years ago, when the groundwater reserves were larger than today, Kally (1973) came to the following conclusion: "The reserve kept in the basins of the system ... for times of emergency is not great and cannot insure us against any possible drought. The recurrence of a prolonged drought, such as the one that visited us for two years running in 1959/60 and 1960/61, will be enough to gravely dislocate the country's water economy" (Kally, 1973). It seems, therefore, that groundwater reserves, even if the current deficit is reversed, could only supply some 2,000 mcm of the required 7,000 mcm in the suggested severe drought scenario.

Desalinated Seawater for the Domestic and Industrial Sectors

Alternative water resources have to be tapped as a substitute in order to reverse the current groundwater deficit and to build up some 2,000 mcm of groundwater reserves as a buffer for times of severe drought, which will also help preserve water quality. In view of the amounts of water required in the current crisis, as well as for future drought mitigation, there is simply no alternative source but the desalination of seawater. It seems that groundwater storage could supply the country with water for only one year on the basis of present consumption, assuming that about 2,000 mcm of groundwater reserves would have been recharged in the basins before a severe meteorological drought would strike. If the drought would last for three or four years, taken as a scenario based on ancient literary data, where would the water come from for the second, third, and fourth year of the drought, after groundwater reserves would have been depleted?

Various economic considerations (Sadan, 1969) have so far acted as constraints in the building of large desalination plants along Israel's coast. However, in view of the current crisis, future water demands, and the need for drought preparedness, financial arguments should have lost much of their clout. In matters of security, large investments are deemed necessary despite the tremendous costs. Water today is also a matter of security and should receive top priority in budget considerations. A new thinking is required concerning desalinated seawater in Israel. If people are prepared to pay the equivalent of about one dollar for a liter of soft drink, they must become used to the idea of paying some two dollars for a thousand liters of fresh, desalinated seawater. All the fresh water required for the domestic and industrial sectors should be derived from desalinated seawater. There seems no other realistic alternative under current climatic conditions and water demands.

A policy to supply all the water requirements of the domestic and industrial sectors from desalinated seawater has a number of great advantages:

1. Annual water consumption is likely to rise with a growing population from the present 450 mcm to 700 mcm by the mid-1990s. This water is absolutely necessary and has to come from someplace.

2. The domestic and industrial sectors are in principle able to pay for the water at a cost of about US$2.00 per 1,000 liters. A precious and scarce commodity ought to have a corresponding price tag, reflecting its real cost. People should get used to the idea that their monthly water bill may become higher than their respective electricity and telephone bills. Water is more important than either of the other two.

3. Israel already has some major sewage treatment facilities. A further expansion in this sector, also necessary to combat pollution and disease and preserve water quality in the aquifers, will enable desalinated seawater to be used twice: (a) first for the domestic sector; (b) then as reclaimed sewage for the agricultural sector. Thus real and significant cutbacks can be made in the amount of water pumped from the groundwater aquifers, so that reserves can be built up for drought years.

4. The desalination of seawater is unaffected by meteorological drought and constitutes as such a very important mitigation factor in drought planning. Even in case of an extremely severe meteorological drought in the southern Levant for a duration of three to four years, as occurred in the past, the supply of desalinated seawater for the domestic and industrial sectors can continue unabated, and its purified waste water can be used for agriculture as well. Thus, the desalination of seawater on a large scale ought to be a key factor in planning for drought in Israel.

Nationwide Treatment of Sewage Water

The preservation of groundwater quality is another extremely important aspect in water resource management and drought planning. Polluted water is not only a health risk, it also threatens the resource of water at its very source: in the groundwater storage basins (Grinwald and Bibas, 1989; Melloul and Bibas, 1990). The improvement of the sewage network in Israel has to be developed, therefore, in conjunction with overall water resource management and drought planning. Every town and village ought to have proper sewage networks and treatment facilities, so that (a) the sewage water cannot pollute groundwater, and (b) the treated sewage water can be used again for agriculture. Besides the use of sewage water from the greater Tel Aviv area, treated domestic waste water from Beer-Sheva has also been used successfully in the Negev desert in sophisticated sprinkler and trickle irrigation systems. Production of cotton, wheat, alfalfa, and corn gave very high yields without any additional fertilizer (Oron and DeMalach, 1987).

The cooperation between the little Negev villages of Midreshet Ben-Gurion and Kibbutz Sede Boker on waste water treatment and its use for agriculture shows that even on a small scale, workable schemes can be developed and executed.

The Mining and Desalination of Brackish Groundwater in the Negev Desert

This water, which is mainly fossil and derived from a wetter climate in the past (Issar and Bruins, 1983), is stored in the so-called Nubian sandstone aquifer. Its present use amounts to about 30 mcm per year (Issar et al., 1972; Issar, 1990). According to calculations by Issar, who has consistently advocated a more extensive use of the Nubian sandstone aquifer, about 200 mcm can be drawn from this aquifer on a long-term basis (Issar, 1990). Some of the water is only slightly brackish and is already used successfully in sophisticated irrigation agriculture. However, the more brackish part would require desalination, which, because of the comparatively lower salt content, will be cheaper than the desalination of seawater. The Nubian sandstone water constitutes without any doubt an important resource that should be included on a larger scale for the further development of the Negev, and as a reserve for severe drought years.

Rainwater Harvesting

The capture and storage of runoff water was an ancient practice in the Near East. Runoff-collecting cisterns were an important source of water for domestic use well into the present century. However, piped water from integrated water sources has replaced the use of local runoff water in Israel. Thousands of remnants of cisterns exist in the Negev Desert from ancient civilizations based on the use of runoff rainwater (Evenari et al., 1971 and 1982; Bruins, 1986 and 1990a). Modern research on the technique of rainwater harvesting in the Negev has demonstrated its usefulness, particularly for developing countries (Evenari et al, 1982; Bruins, 1990b; Bruins et al., 1986 and 1987). The cheap subsidized price of water for agriculture and the relatively large amounts allocated in the last decades have probably been the chief reasons that local runoff water has been largely neglected in Israel. Notable exceptions of the use of runoff in agricultural production are Kibbutz Sede Boker in the arid Negev, and Moshav Kfar Yehoshua with other cooperative settlements in the subhumid Yizreel Valley.[1]

The present water crisis, as well as the need to restrict groundwater pumping, ought to be sufficient reason to stimulate the development of rainwater harvesting all over the country, wherever possible and feasible. Much of this water would otherwise be lost to the atmosphere or the sea. As a matter of fact, interception of surface runoff in local reservoirs has expanded rapidly since 1987 (Grinwald and Bibas, 1989). Runoff rainwater may be intercepted from natural catchments and greenhouses and subsequently stored (Bruins et al., 1991). Storage of runoff water can be achieved in a number of different manners: (1) in the soil profile (as practiced in Kibbutz Sede Boker), (2) in open storage basins (as practiced in Moshav Kfar Yehoshua), (3) in closed surface basins, or (4) in groundwater basins. The latter practice is in fact artificial recharge, which has been carried out by the authorities in the framework of the National Water Carrier system (Grinwald and Bibas, 1989). Rising water prices and smaller water allocations will probably make it more attractive for small settlements to invest more in the interception and storage of local surface runoff.

[1]A *kibbutz* and a *moshav* are different forms of cooperative settlements.

Import of Water from the Nile

In ancient times during severe periods of drought in the southern Levant, people went to Egypt, because the hydraulic Nile civilization gave Egypt immunity from local meteorological drought. With modern technology, Nile water might be transported to the southern Levant. Following the peace treaty with Egypt, importing Nile water might indeed be considered an option, particularly for use in the Gaza Strip and the Negev. The conveyance of Nile water to the east (Figure 2) and the possible large-scale integration of local water resources in the framework of a regional peace treaty has been discussed extensively by Kally (1989). The considerable overpumping of the Coastal Aquifer in the Gaza Strip, constituting its only freshwater resource basin, and the accompanying deterioration of its water quality require a solution in the short term (Schwarz, 1982; Bruins et al., 1991). The situation in the densely populated Gaza Strip is a reflection of the severity of the water crisis in the region at large, because the state of the Coastal Aquifer north of the green line (Figure 2) is almost as bad.

Since Egypt has already built a pipeline to transport water from the Nile to northern Sinai (Figure 2, Sinai Canal), an extension into the Gaza Strip and the Negev might be considered an option to sell Nile water on a commercial basis. A rough estimate of the costs suggests a possible price for Nile water to be in the range of US$0.75 to 1.30 per m^3 (Bruins et al., 1991), which is considerably cheaper than the desalination of seawater (about US$1.50 to 2.00 per m^3). However, the desalination of local brackish groundwater may be cheaper still, about US$0.60 per m^3, although the actual amounts and long-term availability within the Gaza Strip seem limited, and the degree of salinity may increase continuously, affecting the price negatively.

SUMMARY AND RECOMMENDATIONS

Severe meteorological drought is bound to occur in Israel and the Levant in the future, although no accurate drought prediction is possible by deterministic methods because of the complexity of geophysical systems. Proxy paleoclimatic indicators may be useful to reconstruct the main trends of climatic variations in the past, but they usually lack the resolution required to determine the history and impact of annual and multi-annual drought. Though the instrumented meteorological record is of a very detailed nature, its timespan is too short to provide a comprehensive picture of the range of droughts that might be expected in a certain area.

Ancient literary data, therefore, are extremely valuable, because they may provide information not available from other sources about the timing, amplitude, and impact of severe droughts in the past. The Near East is unique in its set of ancient literary data, particularly with regard to the books of the Bible. Severe droughts with a duration of three to four years have occurred in Israel in the past, as shown by the biblical data. Such severe droughts can recur in the near future and are recommended as a scenario for drought-planning purposes.

Groundwater reserves should be the basic ingredient of any drought planning in Israel. Unfortunately, overpumping during the last decades has resulted in a manmade drought with a groundwater deficit of about 1,600 mcm (State Comptroller, 1990). Water resource management has to change in order to preserve water quality and refill the overdrafted aquifers. A groundwater reserve of some 1,600 mcm should be formed. However, in case of a severe drought of three to four years' duration, such a reserve alone would merely be sufficient to supply the water needs of the country for just one year. Other water sources need to be made available, therefore, as soon as possible.

The sea is in fact the only remaining renewable source of water at Israel's disposal that has not yet been used seriously. Desalination of seawater on a large scale in order to supply water to the entire domestic sector and much of the industrial sector should be the main element in new water resource management. Such a policy would allow the aquifers to be recharged without cutting back water allocations to agriculture too severely. The relatively high price of desalinated seawater of about US$2.00 per 1,000 liters should not be a serious constraint for domestic use. Monthly water costs would merely rise to the level of telephone and electricity bills.

Supply of water from a virtually unlimited source, not affected by meteorological drought, constitutes the great advantage of desalination of seawater in terms of drought planning. Sewage treatment of the desalinated seawater, after its use by the domestic and industrial sectors, would also make it partly available as a new water source for agriculture. The comprehensive treatment of sewage water throughout the country ought to be an integral part of national drought planning and water quality preservation. Desalination of brackish water in the Negev Desert and rainwater harvesting throughout the country also constitute significant elements to be included in comprehensive water resource management and drought planning. Higher water prices and smaller water allocations will make both options more attractive than in the past. In addition, the import of Nile water from Egypt might be seriously considered as another possible water source that is not affected by drought over the Levant. Egypt sells oil to Israel, so why not water as well?

Proper drought planning clearly involves much more than advice on drought mitigation, however useful such information may be in its own sphere. A grand strategy is required for sustainable resource management, in which drought risk is seriously taken into consideration. It seems that only comprehensive drought planning on a national level, well embedded in the legislative, executive, and judiciary branches of government, has any chance of being effective, because of the complexity of the problems involved.

Constructive ideas are important, but they will remain on paper unless they enter into the realm of national policy for actual execution. In view of the very serious water crisis in the country, and the added risk of severe meteorological drought in the near future, an independent drought planning task force should be formed. Such a body should be composed of both experts and decision makers to develop detailed plans and enhance actual implementation on a national, regional, and local level. The proposed interaction between participating scientists and decisionmakers will lead to a comprehensive national drought planning effort. Such an endeavor will also help to shape new ideas and solutions that can be implemented by government. Improved drought planning needs to be feasible from a technical-academic point of view and also suited for incorporation into the fabric of government and its institutions.

The generic drought planning process developed by Wilhite and Easterling (1987) and later modified by Wilhite (1991 and 1993) might serve as an initial guideline to develop proper national drought planning in Israel, which seems urgently required. Action is necessary to extricate the country from the current water crisis and prepare detailed proactive drought plans for the future.

ABOUT THE AUTHOR

Hendrik J. Bruins studied earth sciences at the Agricultural University of Wageningen and the Hebrew University of Jerusalem, where he obtained his M.Sc. degree cum laude. He received his Ph.D. degree in 1986 from the Agricultural University of Wageningen. Dr. Bruins is affiliated with the Jacob Blaustein Institute for Desert Research, and he teaches at the Ben-Gurion University of the Negev in the Geography Department. He also practices as a consultant. His expertise includes drought planning, rainwater harvesting, and Third World development. The problems of agro-pastoral societies and food security in arid regions are some of his special interests.

REFERENCES

Begin, Z. B.; W. Broecker; B. Buchbinder; Y. Druckman; A. Kaufman; M. Margaritz; and D. Neev. 1985. *Dead Sea and Lake Lisan Levels in the Last 30,000 Years, a Preliminary Report*. Report GSI/29/85, Geological Survey of Israel, Jerusalem.

Becker, N. 1987. Economic analysis and the testing of variables for a program of drought compensation in Israel. *Hassadeh* 67 (7):1280-84 (in Hebrew).

Bigman, D.; N. Becker; and H. Barak. 1987. Wheat production in semiarid areas and the role of drought compensation: The Israeli experience in the Northern Negev. Department of Agricultural and Resource Economics, University of California, Berkeley, California, U.S.A.

Bruins, H. J. 1986. *Desert Environment and Agriculture in the Central Negev and Kadesh-Barnea during Historical Times*. Published Ph.D. Thesis, Agricultural University of Wageningen, The Netherlands; Midbar Foundation, Nijkerk, The Netherlands.

Bruins, H. J. 1990a. The impact of man and climate on the Central Negev and Northeastern Sinai deserts during the Late Holocene. *In* S. Bottema, G. Entjes-Nieborg, and W. van Zeist, eds. *Man's Role in the Shaping of the Eastern Mediterranean Landscape*; pp. 87-99. A. A. Balkema, Rotterdam, The Netherlands.

Bruins, H. J. 1990b. Book Review: Water Harvesting for Plant Production. *Applied Geography* 10:359.

Bruins, H. J. 1992. Dating of climate and human history in the Southern Levant since the Late Chalcolithic: A basis for climate impact assessment (with emphasis on the Central Negev desert). *In* O. Bar-Yosef and R. Kra, eds. *Late Quaternary Chronology and Paleoclimates of the Eastern Mediterranean* (in press).

Bruins, H. J.; M. Evenari; and U. Nessler. 1986. Rainwater-harvesting agriculture for food production in arid zones: The challenge of the African famine. *Applied Geography* 6:13-32.

Bruins, H. J.; M. Evenari; and A. Rogel. 1987. Run-off farming management and climate. *In* L. Berkofsky and M. G. Wurtele, eds. *Progress in Desert Research*; pp. 3-14. Rowman and Littlefield, Totowa, New Jersey, U.S.A.

Bruins, H. J.; and W. G. Mook. 1989. The need for a calibrated radiocarbon chronology of Near Eastern archaeology. *Radiocarbon* 31(3):1019-29.

Bruins, H. J.; A. Tuinhof; and R. Keller. 1991. *Water in the Gaza Strip: Report.* Government of The Netherlands, Ministry of Foreign Affairs, Directorate General for International Cooperation.

Evenari, M.; L. Shanan; and N. H. Tadmor. 1982. *The Negev: The Challenge of a Desert.* Second edition. Harvard University Press, Cambridge, Massachusetts, U.S.A.

Frumkin, A.; M. Margaritz; I. Carmi; and I. Zak. 1991. The Holocene climatic record of the salt caves of Mount Sedom, Israel. *The Holocene* 1:190-200.

Gilead, D.; and Y. Bachmat. 1973. Israel's groundwater basins. *In* J. Bonné, S. Grossman-Pines; and Z. Grinwald, eds. *Water in Israel—Part A, Selected Articles*; pp. 37-51. Ministry of Agriculture, Water Commission, Water Allocation Department, Tel Aviv.

Gillette, H. P. 1950. A creeping drought under way. *Water and Sewage Works*, March, pp. 104-95.

Grinwald, Z.; and M. Bibas. 1989. *Water in Israel.* Ministry of Agriculture, Water Commission, Water Allocation Department, Tel Aviv.

Hellman, Z.; and D. Rudge. 1991. State Comptroller's report says 25 years of irresponsible mismanagement to blame. *The Jerusalem Post*, January 3, 1991, p. 1.

Hoyt, W. G. 1942. Droughts. *In* O. E. Meinzer, ed. *Hydrology*; p. 579. Dover Publications, New York.

Issar, A. S. 1968. The Pleistocene geology of the Central Coastal Plain of Israel. *Israel Journal of Earth Sciences* 17:16-29.

Issar, A. S. 1990. Climatic changes in the Levant and the possibility of their mitigation. *In* R. Paepe, R. W. Fairbridge, and S. Jelgersma, eds. *Greenhouse Effect, Sea Level and Drought*; pp. 565-74. NATO ASI Series, C-325. Kluwer Academic Publishers, Dordrecht, The Netherlands.

Issar, A. S.; A. Bein; and A. Michaeli. 1972. On the ancient water of the upper Nubian Sandstone Aqufer in Central Sinai and Southern Israel. *Journal of Hydrology* 17 (4):353-74.

Issar, A. S.; and H. J. Bruins. 1983. Special climatological conditions in the deserts of Sinai and the Negev during the latest Pleistocene. *Palaeogeography, Palaeoclimatology, Palaeoecology* 43:63-72.

Kally, E. 1973. Israel's water economy and its problems in the early seventies. *In Israel's Water Economy*; pp. 21-36. Reprints of published papers. Ministry of Agriculture, Water Commission, Tel Aviv.

Kally, E. 1989. A Middle East water plan under peace. *In* H. Ben-Shahar, G. Fishelson, and S. Hirsch, eds. *Economic Cooperation and Middle East Peace*; pp. 48-81. Weidenfeld and Nicolson, London.

Katsnelson, J. 1964. The variability of annual precipitation in Palestine. *Archiv für Meteorologie, Geophysik und Bioklimatologie, Serie B*, 13(2):163-72.

Klein, C. 1961. *On the Fluctuations of the Level of the Dead Sea since the Beginning of the 19th Century*. Hydrological Paper No. 7. Ministry of Agriculture, Hydrological Service, Jerusalem.

Klein, C. 1981. The influence of rainfall over the catchment area on the fluctuations of the level of the Dead Sea since the 12th century. *Israel Meteorological Research Papers* 3:29-57.

Klein, C. 1982. Morphological evidence of lake level changes, western shore of the Dead Sea. *Israel Journal of Earth Sciences* 31:67-94.

Klein, C. 1986. Fluctuations of the level of the Dead Sea and climatic fluctuations in Erez-Israel during historical times. Ph.D. Dissertation, The Hebrew University of Jerusalem.

Melloul, A.; and M. Bibas. 1990. *General and Regional Hydrological Situation in the Coastal Plain Aquifer of Israel and Water Distribution According to Quality Standards (Chlorides and Nitrates) in 1987/88 and Expected to 1992*. Ministry of Agriculture, Water Commission, Hydrological Service. Rep Hydro 1990/3, Jerusalem (in Hebrew, abstract in English).

Ministry of Agriculture. 1973. *Israel's Water Economy*. Reprints of published papers. Ministry of Agriculture, Water Commission, Tel Aviv.

Neev, D.; and K. O. Emery. 1967. *The Dead Sea. Depositional Processes and Environments of Evaporites*. Bulletin No. 41. Geological Survey of Israel, Jerusalem.

Oron, G.; and J. DeMalach. 1987. Reuse of domestic wastewater for irrigation in arid zones: A case study. *Water Resources Bulletin* 23 (5):777-83.

Parry, M. L.; and T. R. Carter. 1987. Climate impact assessment: Review of some approaches. *In* D. A. Wilhite and W. E. Easterling, eds. *Planning for Drought: Toward a Reduction of Societal Vulnerability*; pp. 165-87. Westview Press, Boulder, Colorado, U.S.A.

Riebsame, W. E. 1988. Assessing the Social Implications of Climate Fluctuations. Department of Geography, University of Colorado, Boulder, and United Nations Environment Program.

Sadan. 1969. *An Estimation of Returns to Irrigation Water in Israel with Reference to Seawater Desalting Proposals.* Report No. 305, Giannini Foundation for Agricultural Economics, California Agricultural Experiment Station, University of California.

Schwarz, J. 1982. Water resources in Judea, Samaria and the Gaza Strip. *In* D. J. Elazar, ed. *Judea, Samaria, and Gaza: Views on the Present and Future*; pp. 81-100. American Enterprise Institute for Public Policy Research, Washington, D.C.

Sperber, D. 1978. *Roman Palestine 200-400—The Land: Crisis and Change in Agrarian Society as Reflected in Rabbinic Sources.* Bar-Ilan University, Ramat Gan.

State Comptroller. 1990. *Report on the Management of Water Resources in Israel.* State Comptroller, Jerusalem.

UNESCO. 1979. *Map of the World Distribution of Arid Regions: Explanatory Note.* Man and the Biosphere (MAB) Technical Notes 7. UNESCO, Paris.

Wilhite, D. A. 1986. Drought policy in the U.S. and Australia: A comparative analysis. *Water Resources Bulletin* 22:425-38.

Wilhite, D. A.; and M. H. Glantz. 1985. Understanding the drought phenomenon: The role of definitions. *Water International* 10:111-20.

Wilhite, D. A.; and D. A. Wood. 1985. Planning for drought: The role of state government. *Water Resources Bulletin* 21:31-38.

Wilhite, D. A.; N. J. Rosenberg; and M. H. Glantz. 1986. Improving federal response to drought. *Journal of Climate and Applied Meteorology* 25:332-42.

Wilhite, D. A.; and W. E. Easterling. 1987. Drought policy: Toward a plan of action. *In* D. A. Wilhite and W. E. Easterling, eds. *Planning for Drought: Toward a Reduction of Societal Vulnerability*; pp. 573-89. Westview Press, Boulder, Colorado, U.S.A.

Wilhite, D. A. 1991. Drought planning: A process for state government. *Water Resources Bulletin* 27(1):29-38.

Wilhite, D. A. 1993. *Preparing for Drought: A Guidebook for Developing Countries.* United Nations Environmental Program, Nairobi, Kenya (in press).

World Health Organization. 1988. *Guidelines for Drinking Water Quality.* World Health Organization, Geneva.

Zangvil, A. 1979. Temporal fluctuations of seasonal precipitation in Jerusalem. *Tellus* 31:413-20.

Chapter 9
Indian Drought Management: From Vulnerability to Resilience

A. R. Subbiah*

INTRODUCTION

Numerous strategies have evolved in India to cope with droughts over the years. The characteristics of drought management have evolved from the relief approach of the pre-independence era to the scarcity relief approach after independence. In the 1970s, the scarcity relief approach was replaced by a drought relief approach and, more recently, the drought management approach. These changes in management approaches have increased the country's resilience to droughts. Although India has a comprehensive drought management strategy directed toward preserving the quality of life in communities even in the face of severe and multiyear droughts, vulnerability to droughts at the household level in the dry regions of the country still persists. Hence, evolutionary trends of drought management must continue. It is possible for India to replace its current approach to drought management, which considers drought to be a transitory phenomenon, with a monsoon management approach, which considers drought to be a normal part of climatic variations. This chapter emphasizes the need to evolve appropriate development strategies, keeping in mind the importance of insulating the people of drought-prone areas from the impact of seasonal rainfall fluctuations.

INDIA AND DROUGHTS

Precipitation Characteristics

The dependence of Indian agriculture on monsoon rainfall is well known. For centuries, large-scale variations in climate, particularly rainfall, shaped socioeconomic conditions, cultural ethos, and demographic trends of the Indian subcontinent. The annual rainfall of India is about 117 cm, which is the highest rainfall per unit area in

**Views expressed by the author are his personal views and are not necessarily the views of the organization to which he is attached.*

the world for a country of comparable size. The seasonal distribution of rainfall is: pre-monsoon (March-May), 10.4%; southwest monsoon (June-September), 73.7%; post-monsoon (October-December), 13.3%; and winter (January-February), 2.6% (Sampath, 1989).

The sown area in India can be classified according to the total average annual rainfall as follows: low rainfall region (< 750 mm), 33%; medium rainfall region (750 to < 1,125 mm), 35%; high rainfall region (1,125-2,000 mm), 24%; and very high rainfall region (above 2,000 mm), 8% (Sampath, 1987). This classification indicates that about 68% of the sown area falls in the low and medium rainfall category.

The annual aggregate precipitation of India is about 400 million hectares meters (mham). In addition, it receives 20 mham from countries situated in the Himalayan watershed. This precipitation is distributed as follows: soil moisture, 165 mham; rivers, ponds, and tanks, 135 mham; groundwater, 50 mham; and evaporation, 70 mham. Of the 350 mham accounted for by soil moisture, rivers/ponds/tanks, and groundwater, approximately 200 mham is lost as evapotranspiration. The remainder, 150 mham, flows into the sea or to other countries like Pakistan and Bangladesh. In India, only about 10% of the water provided by precipitation is available, as compared to a figure of 11-12% for the world (Sampath, 1989).

Dry regions having rainfall up to 500 mm per annum are most susceptible to inter- and intra-year rainfall variations. The dry tropical regions of India, where mean rainfall lies between 350 mm and 1,500 mm, are prone to tropical droughts. About 94 million ha, accounting for about 55% of India's total crop area, constitute these dry tropical regions. The vast majority of this area does not have dependable irrigation for even 10% of the crop area. About 300 million people, making up about 33% of India's total population, live in these regions. Vulnerability to drought is aggravated periodically by deficient rainfall. Long-term annual rainfall data over 60 or more years show that more than 13% of India experiences drought once in 2.3 years. More than 50% of the dry tropical region is affected by droughts about once in 4 years (Figure 1).

FROM FAMINE RELIEF TO DROUGHT MANAGEMENT

Drought relief has been in vogue in India for many years. The responses to droughts by the rulers of pre-British India are not well documented. However, the information available shows that no systematic efforts were made to deal with the situations created by droughts. The British administration, particularly in the late nineteenth century, made sporadic attempts to deal with famines by adopting famine codes on the basis of recommendations of the Indian Famine Commission (1880), to be followed by the government when famine threatened lives. The famine codes constituted a milestone in the history of drought management in India: this was the first time a systematic approach, based on climatic, economic, and social factors, was used to guide response efforts.

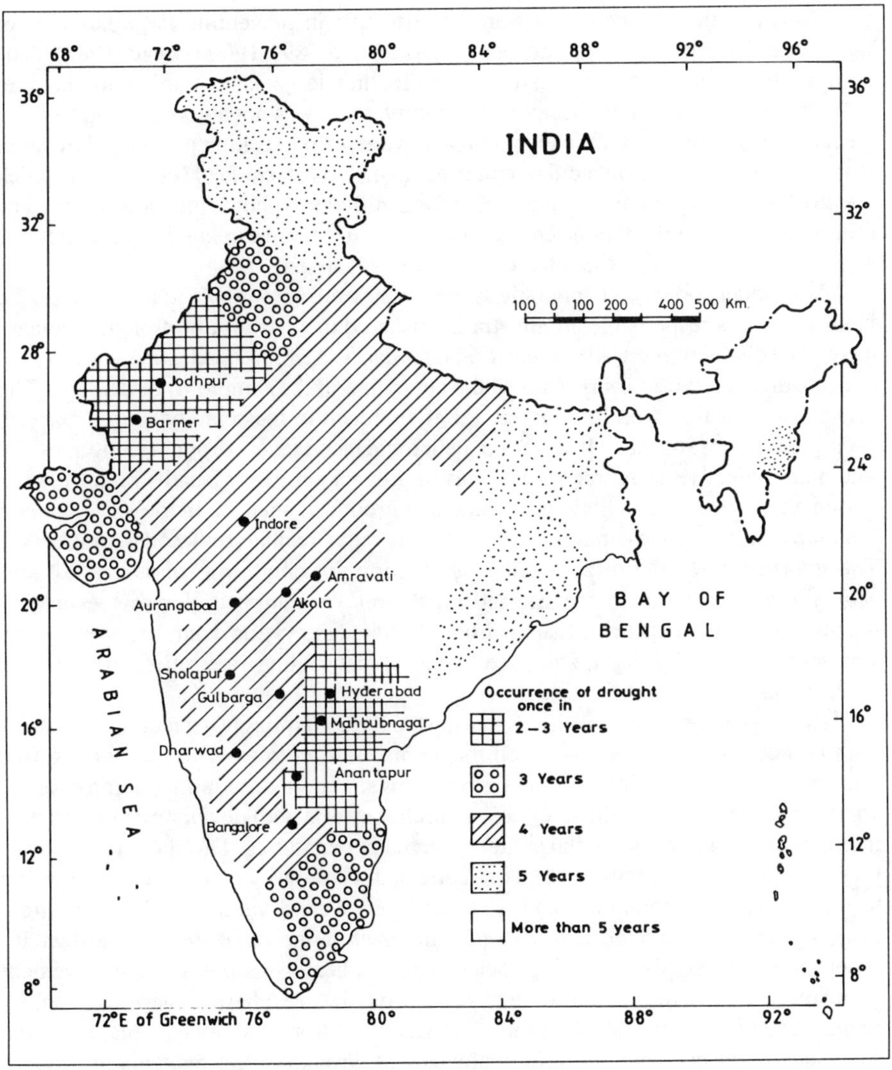

Figure 1. Distribution of areas according to frequency of drought.

However, the famine codes were not effective in preventing large numbers of famine deaths, particularly in the years 1896-97 and 1899-1900 (Dreze, 1988). The famine codes were not successful in eliminating large-scale famine in the pre-independence era primarily because state intervention was not automatic and because the famine codes were excessively punitive toward people seeking jobs on public relief works. In addition, the state did not interfere in the private trade of food grains, which resulted in an unprecedented price rise, forcing millions of people into destitution. The limited objective of the famine codes was to minimize famine-related deaths, and they were implemented only when distress was well advanced.

After independence, democratic systems that were answerable to a free press, the judiciary, and a responsible administration ushered in a new era of drought management. The distinction between *scarcity* and *famine* was emphasized: *famine* was a state of acute distress, but *scarcity* denoted the beginning of the symptoms of distress. The conscious transition from the famine relief approach to the scarcity relief approach occurred during 1965-66, when independent India faced severe drought (Singh, 1972). The main objective of scarcity relief was to prevent starvation deaths rather than to minimize the death toll. State governments introduced changes in the structure and function of famine codes; in some states, scarcity relief manuals replaced famine codes. The states assumed the responsibility of ensuring not only *physical* access but also *economic* access to food for the affected population. An administration was set up with the responsibility of ensuring that no drought-related starvation deaths occurred. This was accomplished by organizing free kitchens, distributing free relief, and opening relief works.

The scarcity relief approach used during the 1965-67 drought—reacting to drought with ad hoc relief measures—forced the administration to resort to organizing free kitchens and other relief programs. To avoid this situation, the administration evolved an approach that ensured the integration of relief efforts with development programs, thus maximizing developmental content of relief expenditures. Drought preparedness measures (such as the establishment of a national food security system) and monitoring techniques for crop conditions and reservoir levels were developed. Thus, a *drought relief* strategy replaced the *scarcity relief* approach in the early 1970s. The scarcity relief approach essentially relied on social distress indicators such as unusual movement of people and an increasing rate of petty crimes, but the drought relief approach mainly relied on economic distress indicators, such as loss of crops and price increases. Some states have replaced scarcity relief manuals with drought relief manuals or drought handbooks.

The lessons learned from the droughts of the early 1970s revealed the existence of administrative, financial, and logistic constraints to organizing relief works for people in local drought-affected areas. During the 1971-73 drought, Maharashtra State employed more than 5 million people on relief projects (Dreze, 1988). The administration found it difficult to cope with the problem of organizing massive relief programs, through which employment was provided to the drought-affected population. As agricultural operations failed because of drought, a large number of farmers and laborers who had been seeking jobs were forced to turn to relief works. It then became necessary to develop a crop stabilization strategy to keep farm labor on the farm operation itself, thus lessening the burden on the state to operate a large number of relief projects. This approach also ensured that food was available at the farm level, thus preventing an overburdening of national food security work. Hence, comprehensive contingency crop planning, based on the experiences of earlier droughts, was formulated.

A three-pronged stabilization strategy (i.e., crop life-saving techniques during the monsoon season, introduction of alternate crop plans based on short-duration crops, and crop compensatory programs to grow crops in favorable areas) was developed (Government of Andhra Pradesh, 1981). This strategy became part of the drought management programs of the early 1980s. Thus, the *drought management* approach has slowly replaced the *drought relief* approach. The drought management approach differs from the drought relief approach in the timing of government intervention and the inclusion of several departments in efforts to minimize crop losses. For drought relief programs, the administration initiated steps after the closure of the monsoon, but the drought management strategy necessitated intervention based on agro- and hydro-indicators before the closure of the monsoon—even (in 1987) in the middle of the monsoon period itself. Table 1 highlights the evolution of drought management in India.

India experienced four major droughts after independence, in the years 1965-67, 1972-73, 1979-80, and 1985-88. Although India struggled to manage the 1965-66 drought by importing more than 10 million tons of food grains and relying on external sources for about half of its resources, the country managed to meet the challenge posed by the 1987-88 drought without food imports and with its own resources. An analysis of impact indicators and distress indicators (Table 2a) between 1965-67 and 1987-88 shows an inverse correlation. Although the severity of the 1987-88 drought is greater in terms of rainfall deficiency, and although this drought affected twice the population that the earlier drought did, the distress indicators show that mitigation of drought impacts in 1987 was more successful than corresponding mitigation attempts in 1965.

The differences in impact and distress indicators between 1965 and 1987 can be explained by resilience and also an improved level of drought preparedness and management as an integral part of the country's development programs. Drought has helped accelerate the development process and, in turn, the process has reduced vulnerability to droughts. The management and development indicators (Table 2b) illustrate this changing situation.

Successive major droughts have helped bring about qualitative changes in India's drought management efforts. The drought of 1965-67 spurred the country into carrying out basic changes in its agricultural strategy; this ultimately ushered in the "Green

APPROACH	Famine Relief	Scarcity Relief	Drought Relief	Drought Management
TIME PERIOD	Pre-Independence era	Post-Independence and up to mid-1960s	From late 1960s to mid-1970s	Late 1970s to present
OBJECTIVES	Minimize starvation deaths	Prevent starvation deaths	Ensure economic access to food	Preserve quality of life
TRIGGERS	—	Social distress indicators	Economic distress indicators	Agro-hydro stress indicators
INTERVENTION	Very late	End of harvest season	End of monsoon season	Within the monsoon season
PREPAREDNESS	Absent	Administrative arrangements	Food distribution network	Contingency crop/ fodder/drinking water plans
KEY MANAGEMENT PROGRAMS	Limited intervention to minimize deaths	Free kitchens, gratuitous relief	Employment generation programs	Crop stabilization
ADMINISTRATIVE APPROACH	Famine Codes	Scarcity relief manuals	Drought relief manuals	Drought handbooks
EPISODES	Pre-Independence famine management	1965 to 1967 drought	1972 drought	1979 and 1987 droughts

Table 1. Major elements of drought nanagement in India: A historical perspective.

	1965	1987
A. Impact Indicators		
1. No. of districts with deficient/scanty rainfall	125	263
2. Rainfall deviation	-18	-19
3. Geographical area affected (%)	66.00	63.00
a. Moderate drought	38.30	29.80
b. Severe drought	0.00	17.90
4. Number of years preceding	0	3 years
5. Crop area affected by drought (% of sown area)	53.23	53.66
6. Population affected (in millions)	47.60	285.40
7. Fall in food grain production (%) over:		
a. the previous year	-18.80	-3.50
b. first crop season (kharif)	-18.80	-7.00
c. the winter crop season (rabi)	-17.70	+2.80
B. Distress Indicators		
1. Rise in food grain prices (%)	65.00	10.00
2. Increase in crimes (theft) (%)	39.20	Negligible
3. Sale and mortgage of land, etc. (% household)	5.53	Negligible
4. Average calorie intake per person per day	1,400	2,000
5. Starvation level (% of household whose calorie consumption fell below 500 Kal)	37.00	1.50

Table 2a. Comparison of impact and distress indicators between 1965 and 1987.

Revolution," which made the country self-sufficient in food grain production. The 1972 drought highlighted the need to create massive employment generation programs to provide the rural poor with income. During the 1987 drought, the dominant concern was the preservation of the quality of life, not merely the prevention of starvation. The government concentrated on meeting the food needs of all persons affected by drought, but officials were also concerned with maintaining normal drinking water supplies and adequate health care and meeting normal energy requirements. The drought policy pursued during this period realized the social and economic goals of the welfare state as embodied in the Constitution.

Reducing vulnerability through development programs has progressively strengthened the coping capacity of the population living in the disaster-prone areas. The higher the development status of a society, the lesser their dependence on government support through relief programs. An analysis of labor strength and affected population at their peak levels in selected drought years is given in Table 3.

The percentage of labor employed compared to affected population has been declining over the years. A survey conducted by the National Council of Applied Economic Research (1986) has shown an increase in nonfarm income over the years. The survey revealed that between 1970-71 and 1981-82, nonfarm income in rural households increased from about 23% in 1970-71 to 33% in 1981-82. One can infer

	1965	1987
A. Management Indicators		
1. Food security system		
a. Procurement of food grains (in million tons)	4.03	13.00
b. Quantity distributed through public distribution system (in million tons)	10.05	21.06
c. Number of public distribution outlets	109,881	333,467
d. Fall in level of food availability from previous normal year (gram per day per person)	79.70	7.20
2. Number of people on free kitchen program (in millions)	11.00	Negligible
3. Number of people on relief works (in millions)	3.00	6.20
4. Total expenditure for drought management (Rs. in billions)	7.22	14.33
a. Domestic share	4.05	14.33
b. International aid	3.17	0.00
c. Amount earmarked for relief (approximate)	5.80	2.00
d. Amount earmarked for productive schemes	1.42	12.33
B. Development Indicators		
1. Net area under irrigation (in million hectares)	26.60	41.00
2. Share of agriculture income (in GDP %)	57.00	33.00
3. Share of agro-based industries in the industrial sector	44.00	34.00
4. Share of thermal sources in total power generation	54.00	70.00
5. Persons below poverty line	50.00	26.00

Table 2b. Comparison of management indicators and development indicators between 1965 and 1987.

from the survey that as a result of the increase in nonfarm income as a percentage of the total income of rural households, the vulnerability of the rural population to natural disasters like droughts has been reduced. This has also reduced the affected population's dependence on the government over the years.

CURRENT DROUGHT MANAGEMENT PRACTICES

The basic approach of Indian drought management focuses on the efficient use of water resources through the adoption of water budgeting strategies, minimization of crop losses through implementation of contingency crop planning, reestablishment of entitlements by providing employment to the affected population through public relief works, preservation of assets of the farms through cattle conservation programs, and a recovery strategy to enable the farmers to resume their normal avocations when the next monsoon begins. This basic approach enabled India to pass through one of the most severe droughts in recent times without causing a setback to its socioeconomic development.

Serial No.	Drought Year	Population Affected (in millions)	Peak Level Labor Strength (in millions)	LaborStrength to Affected Population (%)
1	1965-67	4.76	3.2	6.70
2	1972-73	20.00	9.2	4.60
3	1979-80	22.04	6.2	2.81
4	1987-88	28.54	6.3	2.11

Table 3. Levels of employment during droughts.

India's drought management practices include the following components: (1) operation of an early warning system; (2) drought preparedness measures; (3) conservation of water through water budgeting measures; (4) resource management for stabilizing crop production; (5) assurance to the affected population of physical and economic access to food; and (6) preservation of farmers' assets to enable them to recover quickly after the resumption of the normal monsoon.

Meteorological Monitoring

Monsoon behavior from June to September is the single most crucial index of the onset of drought conditions in any state. Most of the drought-prone states have a standard administrative framework to closely monitor weather-related conditions and their impact on the rural economy. By June each year, a cell is set up under the control of the director of relief at the state headquarters to monitor the rainfall figures from every taluka[1] of the state. This cell also maintains day-to-day contact with the Meteorological Department. Every taluka headquarters is provided with rain gauges, and it is the personal responsibility of the supervisory officer to take readings and convey this information daily via telephone to the officers manning the cell. This data is then compiled on a day-to-day basis and submitted to the director of relief and to key secretaries of the state government.

Hydrological Monitoring

A sophisticated system of monitoring the levels in all medium and larger reservoirs in the state has also been established. By the first of June each year, a cell is also activated under the direct control and supervision of the Water Resources Department of the state government. This cell, which operates around the clock, monitors the levels of most

[1]A *taluka* is a subadministrative unit having jurisdiction over about 50 villages.

of the reservoirs, indicating the total quantity of water available in the reservoir, additional inflow of water during the preceding 24 hours, and loss of water, if any, through either evaporation or spill over. The data from throughout the state is relayed to the central control cell by a network of wireless stations. This daily statement is one of the most important inputs available to government in forecasting drought. An analysis of this data enables government to make policy decisions on the amount of water available for irrigation versus drinking water. By carefully studying the data, the government can make a decision to drastically reduce or ban the use of water for irrigation and to conserve almost the entire quantity of water for drinking water. The Central Groundwater Board is maintaining a national network of more than 11,000 observation wells throughout the country to monitor the behavior pattern of groundwater levels. Monitoring groundwater levels enables the administration to assess the water requirements and take remedial action as needed.

Agricultural Monitoring

The Directorate of Agriculture has a well-established system to closely monitor sowing operations in every taluka of the state, down to the village level. This information provides government with data on the area under various types of crops, progress of the agricultural operations, and health and growth of the crops sown (linked to the advance of monsoons). This information provides the government with valuable data on crop failure.

Early Warning—Union Government

The India Meteorological Department provides a long-range forecast of the behavior of the southwest monsoon each year. The Crop Weather Watch Group meets at noon each Monday to analyze the impact of rainfall on crop conditions and the reservoir levels of 47 major reservoirs monitored by the Central Water Commission. In case any abnormal conditions exist, the Crop Weather Watch Group alerts the central government, which is geared to provide the required assistance to the states.

With the above arrangements, the country's administrative machinery is able to respond to the drought situation in August, before the closure of monsoon and before the people feel the impact of crop failure. Thus, the system does not depend on late drought/famine indicators such as the unusual movement of people, rise in level of petty crimes, distress sale of cattle, or mortality/morbidity statistics.

Although state governments are primarily responsible for management of all natural disasters, the central government supplements the efforts of the states. The district-level administration is responsible for the operational management of disasters.

Hence, a standard administrative machinery exists from village to national level; this can be activated to handle emergencies during drought.

Food Security System

The Food Corporation of India (FCI), established in 1965, provides an elaborate infrastructure for procurement, storage, and distribution of food grains from food surplus to deficit areas. While the FCI, which is a central government undertaking, is responsible for making available essential commodities including food, the state governments are responsible for distributing such commodities to consumers through a network of fair price shops. The National Food Security System operates through the public distribution system (PDS) network through 375,000 fair price shops. The National Food Security System, besides making food grains available to the people at reasonable prices fixed by the government, also operates as an anti-inflationary measure by stabilizing open market prices. The National Food Security System can be strengthened to consider contingencies created by drought to ensure availability of food grains to the affected areas in a timely manner.

Projects

The National Rural Employment Program and the various rural infrastructure building development programs offer employment to the rural work force to stabilize incomes. These employment generation schemes are technically evaluated and kept ready in advance so that they can be implemented to provide jobs to the drought-affected population on short notice. These projects are aimed not only at distributing income to the affected population but also at building up the assets of areas to improve their resistance to drought. Although the National Food Distribution System ensures physical access to food, employment generation schemes ensure economic access to food. Thus, the nation's food security system is an integral part of drought preparedness.

Contingency Crop Plan

The government has a contingency crop plan with buffer stocks of seeds to enable farmers to switch to other crops, depending on the behavior of monsoon. This contingency crop plan, which incorporates arrangements for resource input and energy support such as diesel and power supply, is an important part of the drought preparedness strategy; it helps to stabilize farm-level crop production.

Social Security Schemes

The main social security schemes, such as old-age pensions, midday meals, and integrated child development services (ICDS), are in operation throughout India. More than 230,000 ICDS Nutrition Centres exist; these cover 6.2 million children under three years of age, 6.7 million children between three and six years, and 2.7 million pregnant and lactating mothers. The infrastructure facilities can be expanded to cover the vulnerable section of the population during times of drought.

Infrastructure Arrangements

The vast health-core network that already exists can be expanded to take care of health/nutrition problems that result from drought. The Department of Animal Husbandry, Ministry of Agriculture, with a vast network of dairy cooperatives and veterinary care centers, can take care of the needs of the cattle population. The well-established drinking water/public health engineering department with its vast network is responsible for emergency water supply schemes. The Calamity Relief Fund (CRF) provides more than Rs. 8.0 billion per year to assist in managing emergency situations created by natural disasters, including drought.

Water Conservation

A careful budgeting of water available in reservoirs and the exploitation of surface and groundwater could help meet the critical needs of crops under moisture stress conditions in order to minimize crop loss. During the 1987 drought, the government of India introduced, for the first time, water budgeting to ensure that scarce water resources would be used for optimum benefits. Interdepartmental cooperation at the project/state level was ensured, particularly within the Department of Irrigation. Plans created by various departments and interests (i.e., agriculture, public health, engineering, groundwater reservoir operations, and the State Electricity Board) established drinking water as the first water use priority, followed by fodder programs and irrigation. Although water availability during the 1987 drought was only about 61% of the 1986 level, the available water was sufficient because needs were met on the basis of priority (Figure 2).

Drinking Water

During the 1987 drought, the government of India intensified the efforts of the Accelerated Rural Water Supply (ARWS) Scheme and Technology Mission on Drinking Water. Significant among the measures taken was the release of assistance

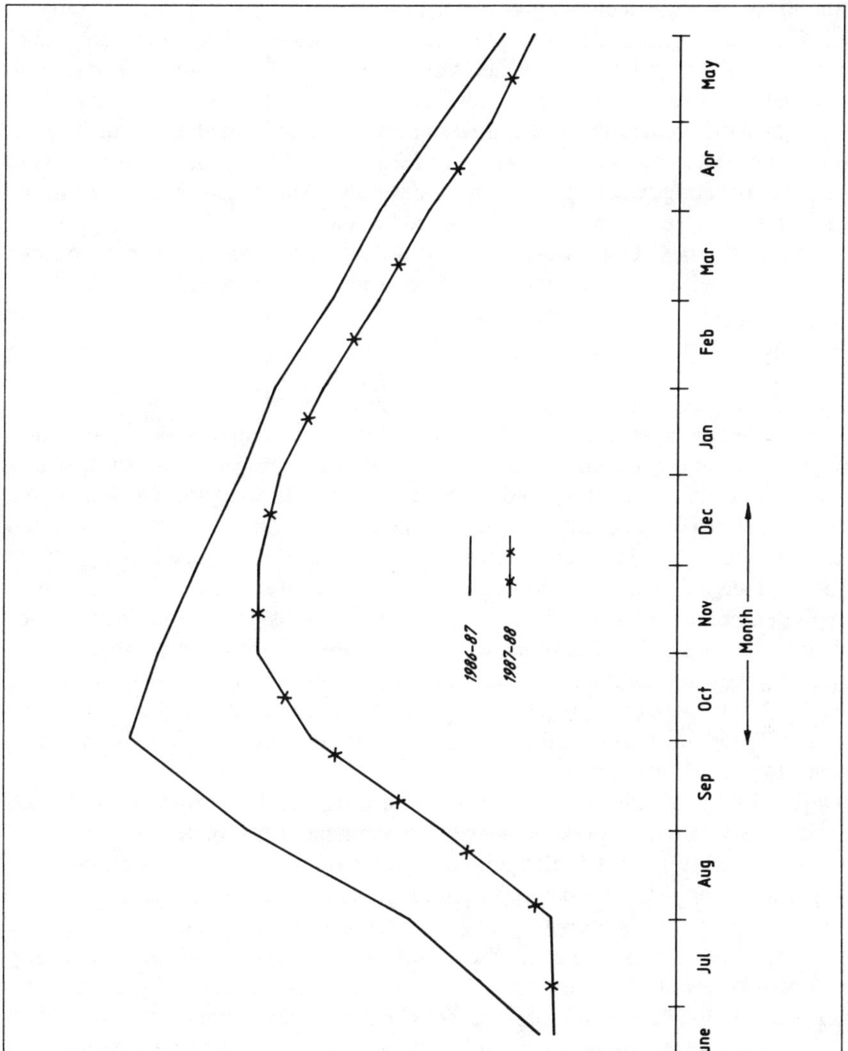

Figure 2. Storage of 47 important reservoirs.

to states to augment the fleet of rigs available for drilling tube wells to provide drinking water. The government also took steps to closely monitor the development and use of 825 rigs available from state governments in the drought-affected areas. The Central Groundwater Board received additional funds to procure 17 high-powered rigs for deployment in different terrains. The government developed a medium-term program to work with 48 towns, with populations of 20,000 to 1 million, that had chronic shortages of drinking water. Major towns like Rajkot, Jamnagar, and Ajmer in the drought-affected areas were provided with additional funds for developing reliable water supply schemes. States were advised to adopt a strict water use plan to manage their reservoirs, with the first priority given to drinking water needs.

Crop Stabilization Strategy

Crop stabilization is ensured through the implementation of contingency crop planning. The contingency crop planning strategy involves three major approaches. First, crop life-saving techniques based on water harvesting and moisture stress-alleviating practices are introduced. Second, programs that provide alternative cropping strategies suited to different weather models are introduced. If it becomes too late to sow rice or sorghum, an alternate crop such as pearl millet, mung bean, or sunflower is recommended, depending on the soil moisture status. Obviously, such alternative cropping strategies will become possible only if seeds of the alternative crops and varieties are available. Therefore, seed reserves are essential for crop security, just as grain reserves are necessary for food security. Finally, measures for improving production in irrigated areas and during nontraditional seasons (i.e., producing alternative crops not traditionally grown during those particular months) through compensatory programs are introduced.

During 1987, the India Meteorological Department and the Crop Weather Watch Group, an interministerial body set up in the Department of Agriculture and Cooperation in the Union Ministry of Agriculture, were able to anticipate the weak monsoon by the beginning of July. On this basis, the central government recommended to the state governments that they mount contingency crop planning efforts. This early recommendation paid rich dividends by reducing the fall in kharif food grain production. The early warning also triggered all-out efforts by different agencies to conserve water resources, budget water, maintain agricultural power supply, ensure the availability of high-speed diesel for agricultural pumps in rural areas, and ensure the availability of seeds.

The crop planning contingencies adopted to manage the 1987 drought not only minimized crop loss during the monsoon season (kharif) but also led to increased production during the post-monsoon season (rabi) in comparison to the previous year. This was the first time that rabi production increased during a drought year.

Year	Percentage Change in CRI from Normal	Percentage Fall in Output over the Previous Year: Kharif	Rabi	Total	Total Food Grains Output
1964-65	+ 6.0	–	–	–	89.0
1965-66	-18.7	-18.7	-17.7	-18.4	72.3
1978-79	+10.00	–	–	–	131.9
1979-80	-20.00	-19.0	-13.7	-17.0	109.7
1985-86	- 4.0	–	–	–	150.5
1986-87	-13.8	- 4.3	- 4.0	- 4.3	144.0
1987-88	-27.5	- 8.3	+ 0.0	- 4.9	137.0

Table 4. Changes in rainfall and food grain output.

According to the cumulative rainfall index (CRI) shown in Table 4, the drought of 1987 and subsequent years was quite severe when compared to previous years. However, the reduction in food grain production was minimized significantly.

Failure of agricultural operations reduces the income of farmers and the labor forces that depend on agriculture. The employment generation program has been the cornerstone of the drought management strategy of India; it provides jobs on public works projects to the drought-affected population. This strategy plays a crucial role in keeping purchasing power in the hands of the affected population in India; it provides economic access to food and enables rural households to preserve productive assets like cattle and land, while minimizing migration. Relief projects such as water conservation, small-scale irrigation, soil conservation, and afforestation programs are intended to impart resilience to the drought-affected areas to better withstand future droughts.

During the 1987 drought, 285 million persons were affected by drought in the various states and Union territories. Nearly 92 million persons belonged to sections of society considered to be vulnerable, mostly small and marginal farmers and agricultural laborers. The immediate impact of the drought was on rural income, especially affecting these vulnerable sections of society. The government found it necessary to provide employment opportunities to these persons on a priority basis. Appreciating the importance of protecting the vulnerable groups, the government earmarked Rs. 8.5 billion for employment generation works. More than 6.3 million persons were employed every day during the peak period of drought in various parts of the country.

To ensure that employment generation works resulted in drought proofing and created durable and productive assets, the government of India had laid down the following priorities for selection of relief projects: (1) tube wells; (2) ponds; (3) field channels; (4) soil conservation and water harvesting works; and (5) road construction, in areas where road links do not exist.

Food Security

The massive public distribution system network had a buffer stock of 23 million tons of food grains at the beginning of the 1987 drought and also was able to procure 13 million tons of food grains from areas with surpluses during the drought. This system effectively ensured the availability of food to the affected population.

Nutritional Security Program

The employment generation program is intended to provide work for the active working population, but it may not ensure enough household income to enable it to provide nutritional support to vulnerable sections of the household, including pregnant women, lactating mothers, and infants. Special nutritional programs are available to take care of the special needs of these sections of the household. Integrated child development services (ICDS) project areas provide special nutrition to these sections of the household during normal times. This project is expanded during drought emergencies to cover more persons. During the 1987 drought, a special nutrition program was launched to provide nutrition support to vulnerable persons who lived outside ICDS project areas. Rs. 5.5 billion was earmarked to cover 7.7 million children and nursing and expectant mothers.

Nutrition Surveillance

The surveys conducted in Rajasthan by the Desert Medical Centre Jodhpur of the Indian Council of Medical Research were analyzed carefully. This analysis showed that in some of the areas, the difference in caloric intake during normal times and drought was often 600 calories. To increase the nutritional status of areas severely affected by drought, a supplementary nutritional program was launched. Under this scheme, additional daily feeding over and above the normal feeding under ICDS and drought nutrition project areas was provided to 700,000 persons in Rajasthan and 47,000 in Gujarat. Thus, nutrition surveillance strategies enabled the administration to take corrective measures during the drought to preserve the quality of life. The nutritional survey (Figure 3) revealed that diets were deficient in calories and vitamin A in all states. However, the survey also showed that during the 1987 drought, not a single household in Gujarat and Andhra Pradesh was subsisting on very low levels of food intake (starvation diet—less than 500 kilo calorie per capita per day). The percentage of such families was only 0.7, 1.0, and 1.5 in the states of Tamil Nadu, Orissa, and Karnataka, respectively.

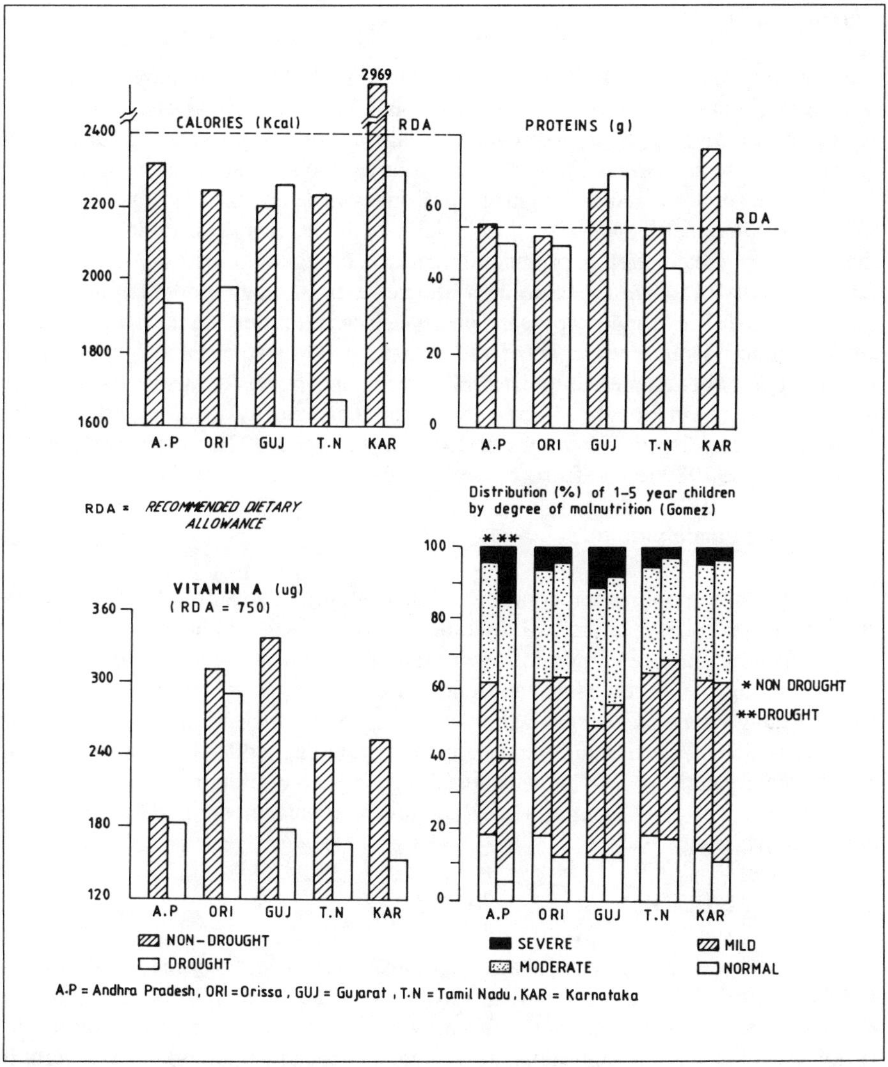

Figure 3. Average consumption of nutrients (cu/day) during drought and nondrought periods.

Preserving Assets

By safeguarding the incomes of vulnerable farmers through employment generation works, it is possible to enable them to retain their valuable assets, such as land. However, it is also necessary to prevent the sale and abandonment of cattle to enable farmers to resume normal economic activities as soon as the onset of the normal monsoon season. Normally, drought results in massive depletion of farm assets through sale and death of cattle, which leads to a delay in the recovery phase. It is important to mount a concerted effort to preserve the cattle in drought-affected areas. Increasing the availability of fodder in the drought-affected areas through fodder depots, giving incentives for raising fodder crops, and preserving cattle in camps or through migration constitutes an essential component of relief programs. Massive transfers of fodder from surplus areas to deficit areas, provision of financial incentives for raising fodder crops on 0.23 million ha in drought-affected areas, maintenance of 1.7 million cattle in 2,258 cattle camps, and distribution of subsidized fodder through 3,422 depots were important elements of the 1989 relief effort.

Community Participation

Community mobilization contributes substantially to social motivation and the effectiveness of relief operations. Providing opportunities for public representatives to contribute to planning and monitoring drought management programs through institutional arrangements at the local and state levels has helped provide relief. Involvement of nongovernmental organizations (NGOs) in the delivery of specific relief to society, particularly in running cattle camps and providing fodder, drinking water, and supplementary nutrition, has helped in reaching the goals of India's relief programs on a large scale. The 1987 drought witnessed massive involvement of NGOs in running cattle camps.

TOWARD MONSOON MANAGEMENT

Region-specific analysis of the drought management strategy reveals that while India's ability to spare and mobilize resources in the drought-prone areas during drought emergencies increased, the farm-level vulnerability in the arid/semiarid dry regions still persists. The physical, economic, and social vulnerability to drought in marginal ecosystems of the arid and semiarid regions is still a main concern of India's development planning.

The massive resource transfers to vulnerable areas that drought entails seldom allow recognition of the inherent capacity of the local systems to cope with drought emergencies (Jodha, 1989). Drought also calls for contingency relief expenditures that consume a significant proportion of each state's budget. Relief expenditures were Rs.

976.9 million during 1979-80 in Rajasthan; the amount increased to Rs. 1.5 billion during 1989-90. To manage the drought of 1987, the government of India earmarked Rs. 14.7 billion for various states. In the long run, the diversion of resources to totally unproductive uses will be detrimental; resources spent on unproductive forms of relief will drain resources needed for the country's developmental needs and, consequently, reduce its ability to manage future droughts.

Employment generation works are undertaken with two objectives: providing relief to the affected population and creating productive assets like small-scale irrigation and moisture conservation works. These projects will enable the drought-affected population to better withstand droughts in the future. During the 1987-88 drought, Rs. 80 billion was earmarked for the employment generation program. Because public works have to be organized within the drought-affected area, numerous and scattered work projects must be organized. In nonirrigated and semiarid zones, drought occurs very frequently and employment must be provided within a stipulated distance of the residents of the affected area. Given the frequency of drought in these zones, it has not been possible for the administration to implement projects such as minor irrigation and soil conservation programs. Because most of the projects are scattered over vast areas involving various sectoral departments, it has not been possible to complete the works. The large number of incomplete works has reduced the productive capability of these projects. As a result, in the semiarid and nonirrigated zones, one of the two objectives of these projects (providing employment to the people to increase purchasing power) is achieved. The other objective (the creation of durable assets) is seldom achieved. Hence, the relief assistance spent through employment generation works in the vulnerable zones has become an instrument of relief provision rather than a measure that creates durable assets, as originally intended.

The depletion of fodder resources forces the administration to procure and transport fodder from distant places to fodder-deficient areas. Similarly, when a drinking water shortage reaches crisis proportions, water is transported by trains and tankers from distant places. The public distribution system also undertakes its massive task of transport and distribution of food grains from surplus areas. Hence, current drought management basically depends on resource transfers in the form of income, food, water, and fodder resources from surplus to deficit areas. This resource transfer approach increases the dependency of these areas and undermines their inherent potential to cope with short-term climatic variation.

Alternate Drought Management Approach

It is critical to define an alternative drought management approach that will gradually eliminate or reduce relief while increasing the capacity of the affected areas to withstand droughts on their own. Drought is a normal feature of climate, and its recurrence is a certainty. However, India's current public drought management practices are designed to respond to drought by treating it as a transient phenomenon, and response efforts therefore are undertaken only for single-season disturbances. Public intervention focuses on the sectoral needs of rural households: the employment generation program is meant for able-bodied persons; supplementary nutrition programs for pregnant women, lactating mothers, and children; gratuitous relief for infirm and physically handicapped people; cattle camps for cattle; and so forth. Contrary to the traditional ways of these communities, the programs treat the family as a single organic unit and attempt to improve the unit's capacity to face the challenges posed by drought. The farming community of the drought-prone regions has developed a diversified occupational pattern to deal with the crisis created by drought situations.

Traditional communities in drought-prone regions have developed short-term and long-term adjustments to handle both good and bad years, but outside intervention artificially isolated drought from the socioeconomic environment of these regions by not recognizing drought as a part of the climatic feature. The farmers of dry regions evolved the monsoon management approach to keep their livelihood system adapted to climatic changes. As such, they manage good, poor, normal, and deficient monsoon years by building up reserves and diversifying economically (Chen, 1991).

Traditional strategies rely on primary, secondary, and tertiary production systems of dry zones to cope with droughts and seasonal fluctuations; these strategies have viable ingredients compatible with the environment. Response systems have mechanisms to absorb the shock of drought and reduce drought impacts. Although the primary production system (such as crops) may fail, secondary production systems (such as livestock) reduce sensitivity. The tertiary production system includes off-farm employment opportunities that provide near-immunity to the impacts of drought.

Because of technical advances, population pressure, and the increasing role of market forces, some of the components of the traditional response system have become redundant. However, the traditional household response system presents a rudimentary monsoon management model, and public policy makers should become more aware of these systems and try to strengthen and revitalize the useful components of these farmer strategies.

The primary and secondary production systems may not be able to absorb the ever-increasing rural labor force in the dry zones. Population pressure puts severe strains on the fragile ecosystems of dry zones, aggravates land degradation, and reduces the natural resource base that supports the livelihood of dry zones. This can result in a reduction in productivity, so it is imperative to reduce the various pressures on the land. The monsoon management model emphasizes the need to facilitate the gradual transfer of large numbers of workers from farm to nonfarm jobs. By enhancing the value of agricultural raw materials, agro-based village industries will create employment and incomes, which will provide income security against droughts. The tertiary production system also can be geared to promoting local skills such as handloom work, handicrafts, and small-scale industries that will provide greater income security.

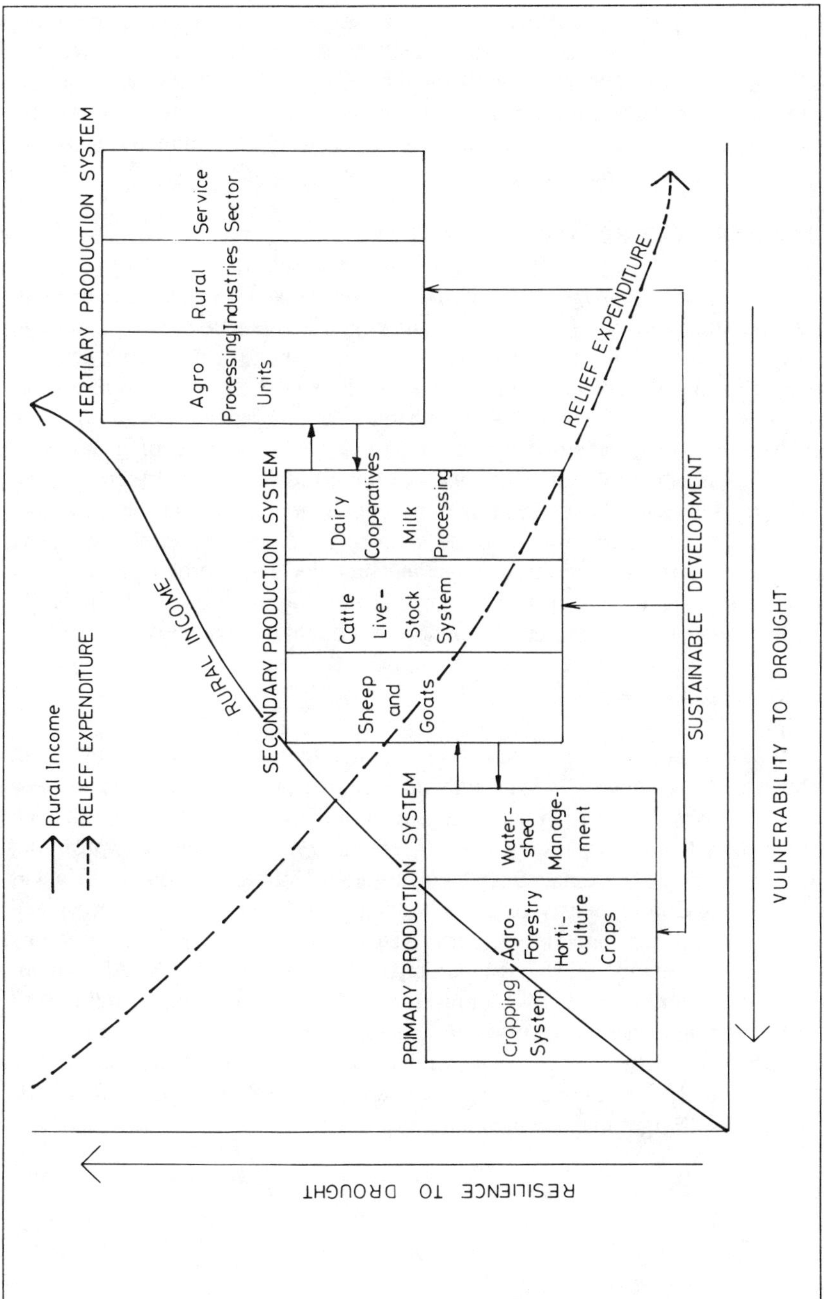

Figure 4. Systems approach to monsoon management.

By adopting a system approach (Figure 4) to monsoon management in upgrading primary and secondary production systems and by building up tertiary production systems, it is possible to transform current drought management methods from resource transfer to resource regeneration. The monsoon management approach is ecologically sound, socially acceptable, and economically beneficial. It will gradually phase out relief expenditure and promote sustainable development.

CONCLUSIONS AND RECOMMENDATIONS

The Indian drought management experience reveals that the current management strategies have stood the test of the time in managing drought emergencies. Although India has increased its ability to mobilize resources to manage drought emergencies, farm-level vulnerability to droughts in dry regions still persists. Hence, it is necessary to initiate a new drought management approach that treats drought as a normal climatic feature. The monsoon management approach, based on the drought-coping strategies of farmers, may be a model that can be used to impart resilience to dry areas. The rudimentary model being adopted by the communities in dry regions may be improved, reoriented, and restructured to incorporate appropriate modern technologies. The approach of learning from traditional management practices and improving them to suit changing societal needs will ensure sustainable development and thus gradually eliminate the need for outside intervention during drought emergencies.

ABOUT THE AUTHOR

A. R. Subbiah, undersecretary to the government of India, Ministry of Agriculture, New Delhi, has been associated with drought management policies and programs since 1985. He was closely involved in the management of the severe drought of 1987 and was associated with the preparation of a document, *1987 Drought Response and Management*, Volume I (national efforts) and Volume II (state experiences). He was a participant in the Comprehensive Disaster Management Training Course organized by the Asian Disaster Preparedness Centre (ADPC), Asian Institute of Technology (AIT), Bangkok, Thailand, in 1987. He participated in the UNDRO/ESCAP Symposium on the International Decade for Natural Disaster Reduction (IDNDR) in Bangkok in February 1991 and assisted in formulating a set of recommendations for drought mitigation for Asia and the Pacific region. Mr. Subbiah has been a consultant to ADPC, Bangkok, on an analytical study of the management of the 1987 drought in India. His area of interest is drought management and mitigation.

REFERENCES

Chen, M. 1991. *Coping with Seasonality and Drought.* Sage Publications India Pvt. Ltd., New Delhi.

Dreze, J. 1988. *Famine Prevention in India.* DEP No. 3, The Development Economics Research Program, London School of Economics, London.

Government of Andhra Pradesh. 1981. *A Handbook for Drought Management.* Revenue Department, Government of Andhra Pradesh, Hyderabad, Andhra Pradesh, India.

Indian Famine Commission. 1880. *Report of Indian Famine Commission 1880.* HMSO, London.

Jodha, N. S.; S. M. Virmani; S. Gadgil; A. K. S. Huda; and B. P. Sing. 1988. The effect of climatic variation on agriculture in dry tropical regions of India. *In* M. L. Parry, T. R. Carter, and N. T. Konijn, eds. *The Impact of Climatic Variations on Agriculture.* Vol. 2, pp. 499-576. Kluwer Academic Publications, Dordrecht, Netherlands.

National Council of Applied Economic Research. 1986. *Changes in the Structure of Household Income and Distribution of Gains in the Rural Household Sector—An All-India Temporal Analysis, 1970-71, 1981-82.* National Council of Applied Economic Research, New Delhi.

Sampath, T. V. 1989. Effect of drought in Indian agriculture. Paper presented at the Inter-Regional Symposium on Ground Water Resources Management in Drought Prone Areas, November 27-December 1, 1989. Central Ground Water Board, Ministry of Water Resources, New Delhi.

Singh, S. K. 1972. *The Indian Famine, 1967.* People's Publishing House, New Delhi.

Chapter 10
Drought and Policy Responses in the Brazilian Northeast

Antonio Rocha Magalhães

INTRODUCTION

This chapter presents an overview of regional impacts of droughts and policy responses in the semiarid northeast region of Brazil (Nordeste). Some ideas discussed here were originally presented by the author at the International Conference on Climatic Impacts on the Environment and Society (CIES), held at Tsukuba University in Japan, 27 January to 1 February 1991 (Magalhães, 1991). The economic, environmental, and social impacts of past and present droughts in the Nordeste; policy responses of the government as well as society; and past, present, and future regional resiliency to droughts are discussed below.

THE DROUGHT-PRONE NORDESTE

The Nordeste covers 1,548.6 thousand km^2, or 18.2% of the total area of Brazil (Figure 1). Depending on how its limits are defined, 60% of this area (roughly 900,000 km^2) is semiarid.

The regional population, as of 1990, was 40 million, representing 27.6% of Brazil's total population. Of this number, 60% (24 million people) live in cities. The rural population is 16 million. The annual rate of population growth, from 1970 to 1980, including migration, was 2.2%. The natural rate (birth rate minus death rate) of

	1980	1990
Urban	17,476.7	23,687.5
Rural	17,171.8	16,083.3
Total	**34,648.5**	**39,770.8**

Table 1. Population of the Nordeste (in thousands).

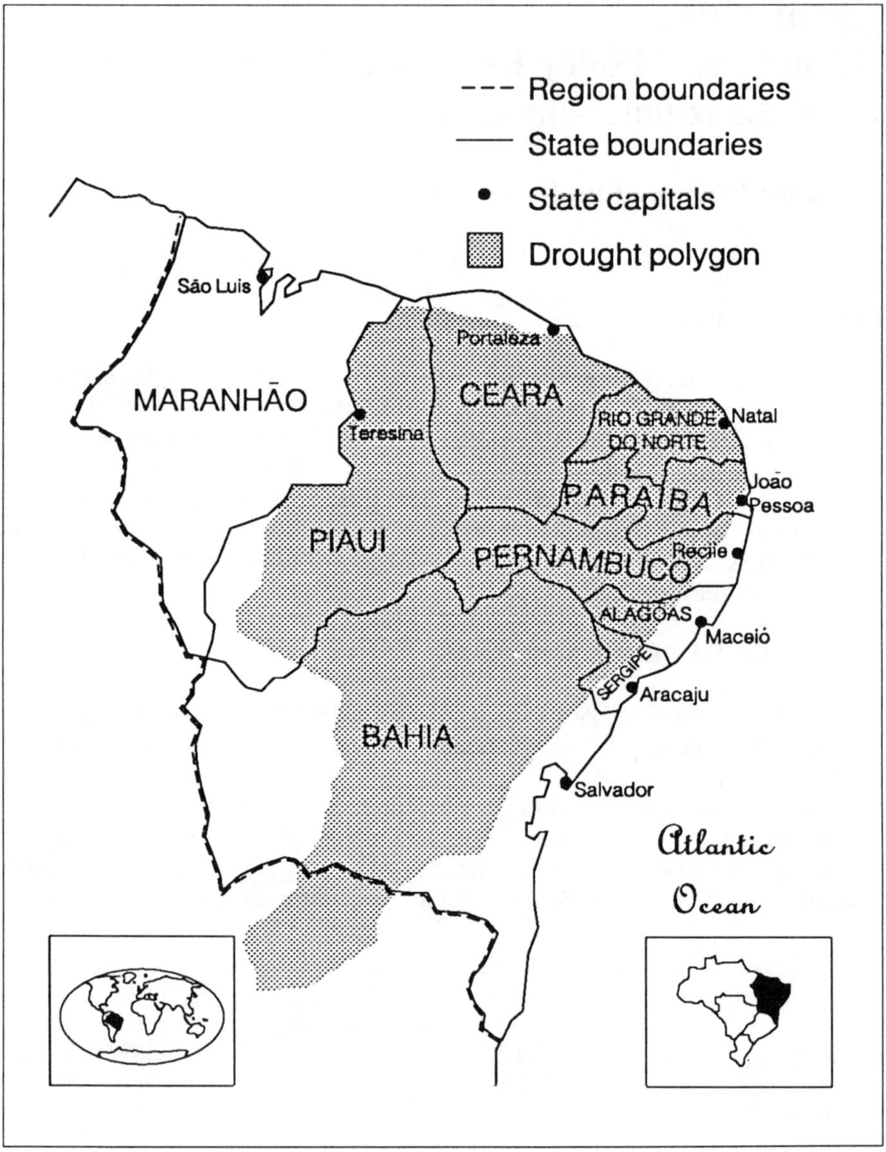

Figure 1. The semiarid northeast region of Brazil.

population growth was 2.8%, but part of the population continues to leave the region. Urban population growth was 4.0%, while the rural rate was 0.53%. These rates now show a downward trend. From 1980 to 1985, the annual growth rate of the regional population was 1.5%. The population growth rate is decreasing, and the absolute size of the rural population is already decreasing (Table 1).

Nine of the twenty-six Brazilian federal states and three of the ten most populous cities in Brazil are in the Nordeste. Historically, the Nordeste is the cradle of the country. For more than 300 years, Salvador was the capital city of Brazil. From the sixteenth to the eighteenth century, it was also the main economic region, with sugar cane production and trade attracting the attention of Portugal and other nations of Europe, while livestock, cotton, and subsistence crops ensured the occupation of the semiarid hinterland.

Though the Nordeste is not a political unity, it is broadly recognized that some physical, cultural, economic, and institutional characteristics exist that create a regional identity. These characteristics include the semiarid environment, economic backwardness and social problems, a common cultural background, and the area's vulnerability to climate fluctuations.

The Semiarid Region

The semiarid region is a heterogeneous area, characterized by semiaridity, xerophilous vegetation, and recurrent droughts. Precipitation ranges from 300 mm/year to 800 mm/year, depending on the time and place. Water resources are scarce. The region has only two permanent rivers: the São Francisco River, which comes from the state of Minas Gerais (southeast region) and crosses the semiarid areas of Bahia, Pernambuco, Alagoas, and Sergipe; and the Parnaiba River, between the states of Piauí and Maranhão. The other rivers of the semiarid region are temporary rivers, because they become dry during the dry season. Underground water is scarce and unevenly distributed. Surface water is more evenly distributed, primarily because it is accumulated in dams that have been built during the last nine decades.

Soils are generally shallow. Though some soils are of good quality, they are laid on crystalline rocks that are often exposed at the surface. Sedimentary soils are present in some areas, especially in river basins; together with other deep soil spots that can be found in several points of the semiarid region, they offer some potential for irrigation. Shallow soils and rockiness severely limit the use of mechanization in many areas.

The vegetation is dominated by xerophilous plants. Although several different ecosystems exist, caatinga vegetation—mainly small and thorny trees, generally sparse and highly diversified—predominates over large areas. Depending on the type of soil and microclimate, the caatinga may become denser and the types of trees may be different. In the driest areas, cactus and thorny bushes are more frequent. In the more humid—or less dry—areas, and on some mountain spots, "islands" of microclimates exist, where temperatures are mild, precipitation is higher, soils are deeper, and trees are tall. Unfortunately, these are only small islands surrounded by the large dry ecosystem.

The Nordeste semiarid region has been occupied since the sixteenth and seventeenth centuries by settlers in search of land for cattle raising, together with cotton production and subsistence food crops. One result of this occupation was a pattern of land use in which large and small land holdings coexist. Human settlement remained sparse and increased slowly during the first centuries. The use of inappropriate technologies contributed to decreasing agricultural productivity levels over the years. Clearly, natural resources of the semiarid region today are overused, with present technologies, practices, and level of use exceeding the region's carrying capacity.

The current population of the semiarid region is about 20 million people, living in small and medium cities and in rural areas. Since the last century, overpopulation was observed only during droughts, but today this is a permanent condition even in "good" years. The population density is clearly in excess of the land's carrying capacity at present levels of technology and economic activity. As a consequence, the population is continuously pushed to emigrate to other parts of the country (see Bitoun et al., 1992). The majority of the outmigrants are the small landholders and landless laborers who are unable to withstand prolonged adverse climatic conditions.

The main economic activities in the semiarid region are still cattle raising on large landholdings and raising food subsistence crops on small landholdings and in sharecropping. The most important cash crop, cotton, has almost been destroyed during the last 10 years by the boll weevil, and this has destroyed the main economic basis of semiarid agriculture and economy. Hence, the rural economy of the Nordeste has become disorganized and new alternatives have not yet appeared. However, looking to the future, irrigated agriculture seems to be the alternative for those areas where irrigation is possible—about 3% of the Nordeste territory.

Landownership distribution is highly concentrated in the Nordeste. According to SUDENE (1985), the smaller landholdings, with less than 10 ha each, represent 34.3% of the number of landholdings and occupy only 1.4% of the area, while the larger ones, with more than 10,000 ha each, represent 0.07% of the number and 22.8% of the area. In 1980, the rural Nordeste had 2,450,000 rural producers; 1,031,000 of them did not own their land. The large majority were small producers. The small farms in the semiarid region produce subsistence crops like beans, maize, manioc, and sweet potato, while the larger ones are mainly dedicated to cattle raising.

Economic Backwardness and Social Problems

Income per capita in the Nordeste is presently around US$1,000 a year, about half of the same figure for Brazil as a whole and a fifth of that of São Paulo, the richest state in the country. Interregional inequalities are pronounced. Although the Nordeste has 28.5% of the country's population, the regional Gross National Product (GNP) represents only 14% of Brazil's GNP. Inequalities within the Nordeste are also significant. Large income differences exist between the urban and rural population, as well as between social groups. Although we have no records of rural income in the semiarid region, one can estimate that it is below US$200 per year.

The majority of the Nordeste population is considered to be below the poverty line. Indeed, 74% of the population 10 years of age or older (defining the economically active population according to the official measures of the Instituto Brasileiro de Geografia e Estatística [IBGE]) have no earnings or earn less than one minimum legal wage per month—about US$60, less than needed to provide for the basic needs of any person. Rural poverty is greater than urban poverty. Indeed, many of the urban poor are illiterate migrants from rural areas. In the rural area, 83% of the population are below the poverty line, compared to 68% in the urban areas.

In the Nordeste, life expectancy is 51.6 years, compared to 60.8 years in Brazil as a whole. In 1980, 36.4% of the population 10 years or older were illiterate or had attended less than one school year, and 74.3% had attended less than four school years. The infant mortality rate was 118.8 per thousand born alive. The rate of emigration in the Nordeste is about 19.5%, compared with a rate of immigration of 6.5%. Net emigration is 13%.

Vulnerability to Climate

The semiarid Nordeste is highly vulnerable to climate variation, especially to frequent droughts. Since the carrying capacity of the area has already been exceeded, any climatic variation that causes a negative impact on production, employment, and water availability may disrupt the fragile social, ecological, and economic systems.

Social vulnerability is the region's main problem. The majority of the population is dependent on the "good behavior" of the climate for their living. "Bad behavior" seriously affects them. The ecosystems, though originally adapted to climate variations, became vulnerable because of inappropriate land use practices. Economic vulnerability is high because of the predominant role of subsistence rainfed low-productivity agriculture practiced by small farmers and landless farmers.

Cultural Background

The common history of the occupation of the semiarid region ("man following the cow"), the isolation in which the population lived for several hundred years, and the continuous struggle with the wild environment forged a culture that is very similar throughout the various states of the Nordeste. An inhabitant of the Sertão, as the semiarid region is known, is called a *Sertanejo*. At the beginning of the twentieth century, a German scientist working in the Sertão wrote:

> The Sertanejo lives in constant fight against the environment that surrounds him; he endures long periods of water scarcity for the subsistence of his family and his cattle; he endures with resignation the calamitous rains that result in heavy floods and, during this time, his scarce food gets spoiled or disappears and his plantations are extinguished. These continuous shortcomings make him stubborn, resistant, and rough (Luetzelburg, 1922-23, p. 105).

Dependence on nature, especially on climate variations; a close relationship with the surrounding environment; a strong religious sentiment that makes him accept as given his destiny and the difficulties of life (and, in a way, the belief that he is not able to change things); a lack of preoccupation about the future, making him worry only about his present-day subsistence; the courage to face his day-to-day difficulties, of which the fight against droughts is the most important; and the disposition to work and never give up—all of these are traces of the culture of the Sertanejo and components of his character. Though the Sertanejo keeps most of these characteristics, there are many indications of change in recent times, due to urbanization, reorganization of the agricultural system, and modern communications and education. In the last three decades, grassroot institutions guided by the Catholic Church and unions of rural workers are helping raise the level of awareness of the poor rural population.

THE REGIONAL CLIMATE

The Nordeste climate is tropical and ranges from moist to dry. Temperature varies along the coastal Atlantic zone from average lows of 20°C to average highs of 30°C. In the semiarid interior, average highs are higher. Temperature variations among seasons are rather small. A continuous wind from the Atlantic Ocean makes the climate seem mild, especially in coastal areas and river valleys, but high temperatures in the Sertão can easily reach 38-40°C.

Year	Drought/Rain	Coef.Variation
1970	Drought	.69
1977	Rain	.32
1983	Drought	.43
1986	Rain	.78

Table 2. Precipitation in the Nordeste: Coefficient of intra-annual variation. Source: Arraes and Castelar, 1989.

Precipitation is highly variable, spatially and temporally, both intra-annual and interannual. The variation coefficient of precipitation may be as high as 78% within the same year (Table 2).

Precipitation figures vary between 300 mm and 800 mm per year, according to location. High intra-annual variability may mean a concentration of rains in a few months or days, causing floods and droughts within the same year. High interannual variability may mean years of good rains, years of excess of rains, and years of drought.

Rains in the Nordeste are influenced by three main regimes: the penetration of the Antarctic cold fronts in the southern Nordeste, the southeast Atlantic fronts in the coastal areas, and the Intertropical Convergence Zone (ITCZ) in the northern Nordeste (Nobre and Molion, 1988). There is evidence of a correlation of droughts in the Nordeste (and floods in the south of Brazil) with the behavior of ENSO (El Niño-Southern Oscillation) events (see, for instance, Gasques and Magalhães, 1987).

Droughts are the main climate problem in the Nordeste, especially in the Sertão. They are a recurrent phenomenon and an ancient problem. Drought frequency is high. In eight of every ten years, there is a chance that some type of drought will occur. Some areas are more drought-prone than others. Droughts are more frequent in the northern part of the Nordeste, but they occur throughout the region.

Floods are also a serious problem, though of a very different nature. Large floods have been recorded a number of times, including 1924 and 1974. At the beginning of the twentieth century, Luetzelburg wrote:

> Working always under an inclement sun, suffering from lack of water, suddenly this picture was transformed, because the rains came heavy during March...The rivers, until then with scarce waters...became torrents; the low lands became extensive lakes, increasing in volume, interrupting completely the transit to other cities. The rivers...in impetuous stream, destroyed everything that was not solid; cotton packages, rice crops, so painfully cultivated by the farmers on their margins, big Jatoba trees with their roots showing up above the waters, roofs of the marginal houses; everything was destroyed (Luetzelburg, 1922-23, p. 93).

Droughts, as well as floods, have been and still are a serious problem. During 1990, the Nordeste suffered again from droughts and floods. In 1992, drought is again affecting the population of the Sertão. Climate variability will certainly continue in the future. If the Intergovernmental Panel on Climate Change (IPCC) report is correct, droughts will become more frequent in semiarid regions of the developing countries, like the Nordeste. The IPCC report includes parts of eastern Brazil, where the Nordeste is located, as one of the world regions most at risk from climate change (IPCC, 1990).

IMPACTS OF CLIMATIC VARIABILITY AND DROUGHTS

Climatic variability imposes a heavy burden on the Nordeste economy and living conditions as well as the regional environment. Agriculture is the most affected sector, because of its dependency on climate conditions. In Table 3, I compare a drought year and a nondrought year, in terms of the main agricultural crops.

	Dry Year-1983	Wet Year-1978
Beans	24.8	120.0
Rice	30.1	67.2
Corn	17.5	259.0
Cassava beans	442.1	1,575.1
Cotton (annual)	17.2	27.7
Cotton (permanent)	46.5	237.0

Table 3. Impacts of climate variability on agriculture in the Nordeste (in 1,000 tons). Source: Fraga and Almeida, 1989.

The impact of drought is usually less severe on cattle raising than on agriculture. Only extreme droughts affect livestock activity seriously. Urban activities are less affected, except those that are directly dependent on agricultural production: some types of agro-industries, the cotton trade, and the export of food crops. Prices of food become higher. However, trade may increase, because of increased movement of imports from other regions (Magalhães and Rebouças, 1988).

Manufacture in general may be affected when droughts outside the Nordeste, on the headwaters of the São Francisco River, cause a fall in hydroelectricity production. In this case, electricity rationing may cause a decrease in industrial output and an increase in industrial unemployment. The 1986 drought in the headwaters of the São Francisco River caused a reduction of 6% in total energy consumption and a fall of 15% in the industrial employment in the Nordeste (see, for instance, Duarte, 1989).

Public finances are affected in two ways: first, droughts cause a fall in tax collection, because of diminished agricultural output. Nevertheless, this decrease may be smaller than the fall in output, because of price increases. Second, if government action injects money into the economy as a response to droughts, a net increase in tax collection may occur because of increased monetization and increased trade. The final result also depends on the state of the national economy, in the sense that an economic recession will imply a decrease in tax collection. If Brazil does not experience an economic recession, a drought in the Nordeste will cause an increase of about 10.4% in state revenues (see Arraes and Castelar, 1989). This happens because "a) a drought causes a fall in agricultural product and hence in total product, but b) triggers federal government relief actions...that c) monetize the economy through cash payment of wages and purchase of inputs, so d) stimulating the services sector activity, mainly civil construction (through the work-fronts), the commerce and the import sector, thus e) increasing the base for state tax collection" (Magalhães and Glantz, 1992, p. 38).

The impacts of droughts on natural resources and the environment are also serious. Drought impact on water resources has been the main problem, because it affects water availability for human and animal consumption. Normally, the semiarid environment is adapted to climate variability and droughts. Excessive land use due to settlements and economic activities has caused heavy environmental stress. Under these circumstances, droughts are a cause for further environmental stress and desertification.

The economic and environmental repercussions of droughts result in serious social problems for the vulnerable population of the semiarid region. Thirst and hunger historically have been associated with droughts. The semiarid population is, for the most part, very poor, and they have no means to withstand drought crises. If agricultural production ceases and drinking water is not available, then subsistence conditions suddenly cannot be maintained, and the population becomes deprived of their means of surviving. In the past, this situation meant massive death. More recently, government responses and infrastructure facilities have been effective in avoiding that disaster, but the population is still extremely vulnerable.

Historically, the drought disasters provided their own kind of adjustment to the semiarid environment, at very high economic and human costs. The following passages were written in 1847:

> [After the 1723-24 drought] things improved with the return of the rains. Livestock multiplied and population increased. In 1744, there was a second drought; in this, all the livestock died and people suffered of hunger to such a degree that malnourished children that already walked began again to crawl.
>
> After 1744, things improved again, and again livestock and human population increased. After 22 years, in 1766, there was a third drought; in this there was also much livestock death and human hunger.
>
> Things improved again and once more livestock and people multiplied. After 13 years a new drought occurred in 1791 and again there was livestock death and human hunger (Guerra and Guerra, 1909, p. 16).

This cycle continues. After an extreme drought year, the times improve and everyone forgets about it. The next drought finds people unprepared, and the impacts are as severe as ever—except mortality rates are no longer so high, as was the case during the big drought of 1877, when approximately 500,000 people died of hunger and thirst.

> The [1877] drought was devastating in all...provinces...in all of them the population was reduced to misery, ruin and poverty; the picture was terrible...the extinction of the moving animals was complete.
>
> The mortality in the last months of the year was astonishing, everywhere (Guerra and Guerra, 1909, p. 39).

During the 1983 drought, 12 million people were directly affected. They lost their jobs and any possibility of earning an income. Their vulnerability was complete. Subsequent droughts in 1987, 1990, and 1992 continue to reveal high levels of social vulnerability.

Government and societal responses to drought have helped eliminate the most calamitous impacts of the droughts, but much needs to be done before the Nordeste population can become resilient to the climate variabilities that affect the region.

POLICY RESPONSES

Three broad categories of responses to droughts are discussed below: (1) responses by the most vulnerable segments of the population (rural workers, sharecroppers, and small farmers); (2) responses by farmers and landholders in general; and (3) government responses.

Responses of Vulnerable Groups. The most vulnerable groups, in the event of an extreme drought, face a problem of survival. Without a source of income, they look for alternatives, such as

- Finding another income-generating activity, although in general no alternatives are available.
- Hunting wild animals, even rats and snakes, and collecting roots and cactus to eat. This was a common response in the past as well.
- Looting storehouses in the neighboring cities, in search of food.
- Digging wells on the dry rivers, in search of water.
- Migrating within the semiarid region, to places where the drought is less severe. This alternative is unavailable in an extreme drought that affects the entire region.
- Migrating to the capital cities of the Nordeste states (all but one of these cities are located in the coastal zone), big cities of Brazil (more Northeasterners live in São Paulo than in any city of the Nordeste), and other regions (the Amazonian region has been populated by Northeasterners fleeing from the droughts) (Bitoun et al., 1992).

- Getting help from government relief programs. In recent times, this has been the main—and sometimes only—alternative, since migration becomes more difficult during periods of macroeconomic recession, as was the case in Brazil in the 1980s and early 1990s, because of the external debt crisis and the need for macroeconomic adjustment.

In times when government relief programs were not available promptly, and possibilities of migration were limited by the lack of roads and transportation, the stricken population faced droughts with very few viable alternatives. Trying to leave their homes without a real destination, they often starved along the roads. In 1845 (and in 1877, 1915, 1919, 1932, 1951, and 1958 as well),

> The poor population left their homes because of hunger, and many died while traveling (on foot), everything lacking, hunger, nakedness, thirst. [Even the] rich people found themselves in great difficulties and many of them faced indigence (Guerra and Guerra, 1909, pp. 27-28).

During present-day droughts, the alternatives available to the stricken population are even more limited because of environmental stress caused by overpopulation and inappropriate land use (i.e., in excess of land-carrying capacity). In 1991,

> Hungry farmers looted the fairs in the neighboring municipalities of Bodocó and Ipubi (600 km from the capital city of Pernambuco State), taking more than 12 tons of corn, beans, and flour. Seven tons of basic food were also looted from a popular food supply agency of the State Government.
>
> The people are hungry, we need food for our women and children. The only way was to take it.
>
> The mayor of Bodocó [said that] what happens here is out of necessity. Nobody has agriculture, nobody has a wage, there is no work front for the workers...The municipality lost more than 90% of the beans and corn (*Jornal do Brasil*, Jan. 9, 1991, p. 5).

The values of present and future (twenty-first century) civilizations include an emphasis on economic, social, and technological progress capable of solving the main problems of society. However, this progress has not been able to solve the problem of the extreme vulnerability of the peoples of semiarid regions, including the Nordeste. Within a world of progress and well-being, the continued existence of the extreme vulnerability of the semiarid Nordeste cannot be justified. Why has modern civilization been unable to solve such a basic problem of survival that affects the majority of the population of the developing regions?

Farmers and Landholders. Long before the rainy season, the farmers, as well as the rural population in general, make an effort to discover signs of rains or droughts. The experiences vary from very simple to more complex and pseudoscientific. Whatever the degree of complexity, they are an important aspect of the culture of the rural populations and their popular faith.

When the drought comes—considered official if rains do not come by March 19, the day of Saint Joseph—farmers react:

- Stopping any dry farming agriculture. As a result, rural workers and sharecroppers are immediately unemployed, and that may mean several million workers unemployed from one day to the next.
- Trying to protect their livestock instead of agricultural activity. First, farmers reserve the available water and the fields for the cattle; second, if this is not enough, they take the cattle to neighboring areas or even neighboring states, where water and pasture is available. This is the reason that livestock is affected only during extreme droughts (see, for instance, Magalhães and Rebouças, 1988, p. 288).
- Selling the cattle to other farmers or slaughtering the herd if no other alternative exists, in which case the price of cattle comes down. This practice is more common in the case of small farmers, and many owners, particularly those with large landholdings, take advantage of this, because after the drought the prices increase sharply.
- Making investments in water resources, building dams, and digging wells on their properties, often with government financing and incentives.
- More recently, adopting irrigation practices.
- Less generally, cultivating xerophilous plants, especially those that provide food for the cattle.
- Feeding wild cactus to cattle.
- Selling their lands and migrating.

Government Responses. Until the end of the last century, government response to the problem of the Nordeste droughts was almost nonexistent:

> The magnitudes of the climatic phenomenon that affects the Nordeste, maybe because of its distance, still did not impress, even sentimentally, the men of government and of science (Braga, 1982, p. 37).

This quotation refers to the year 1859. Of course, there was some sparse and ineffective response. In 1877, the central government (at that time, the Empire of Brazil) spent

60,000 contos (roughly 4 million sterling pounds) to buy food in other regions to distribute to the "scourged population" stricken by the drought.

The 1877 drought was a tremendous disaster. It captured the attention of the nation, and in many parts of the country, committees were formed to help the drought victims. The government formed a technical commission to study the problem. In 1906, the first big public dam was built, and a few years later an institution was created to combat the effects of the droughts: the Inspectoria Federal de Obras Contra as Secas - IFOCS (National Inspectorship for Works Against the Droughts), presently called the National Department for Works Against the Droughts (DNOCS). Other Nordeste institutions have been created in the decades since 1940 (Pessoa, 1987).

During the first half of this century, the major emphasis was on building dams (because *drought* means lack of water) and roads (to enable the population to flee from the droughts). The idea of the "work fronts" (government work projects) was introduced, with the aim of employing the drought-stricken population during droughts to build dams and roads.

Many dams were built as part of the development of water resources in the region. From 1909 to 1983, DNOCS built 275 big dams with an accumulation capacity of 14 billion cubic meters of water (DNOCS, 1985, p. 11). But all this storage capacity was not well distributed. DNOCS has been criticized because most of the water accumulated in dams was not used to foster agricultural production, as in irrigation projects.

More recently, other types of work fronts and relief programs have been practiced (see, for instance, Magalhães et al., 1989), including works in the interior of private landholdings, small community works, community irrigation, land preparation for agriculture for small farmers (in anticipation of the next rainy season), and providing subsidized credit for the stricken population.

Despite their problems, all of these measures have had some positive effects, yet the sentiment remains that the general level of resiliency of the population did not improve. This sentiment has lead the government to be hesitant in defining a permanent policy to face drought problems, and this has sometimes resulted in retrogression. Indeed, the federal government in Brazil has historically responded to the drought problem, but only during droughts. Normally—as in ancient times—the drought is forgotten when it is gone. When it comes back, a new critique of the past drought policies—especially of the problems in using public resources—is made and new strategies are announced. However, these "new" strategies are only repetitions of previous practices.

One of the important improvements in recent strategies was achieved during the 1987 drought, when some states—especially Ceará—adopted a strategy of employing the stricken population in community works and managed to sidestep political influences that traditionally resulted in patronage and paternalism. This new strategy improved the planning, implementation, and evaluation of the works, through a democratization process that involved the participation of the communities and representatives of the local leaderships in the decision-making process.

From the century of experience in policy response to drought in the Nordeste, a number of recommendations can be made:

- For the highly vulnerable segments of the population, relief action (or emergency action) that may assume different forms is needed. Relief actions should be available to provide a steady income for the stricken population (through public works employment, for instance); provide or distribute drinking water; or facilitate the food trade and distribution.
- A permanent development policy should be created that increases the population's income level on a stable and ongoing basis, so that in the long run the region as a whole will become more resilient to drought.

REGIONAL RESILIENCE TO CLIMATE VARIABILITY

In assessing the present resiliency of the Sertão to droughts, two main issues must be stressed:

1. In contrast to the past, some infrastructure (although still not enough) is available to increase regional capacity to face droughts. This infrastructure resulted from building public works during droughts (when the main objective was to offer jobs and income opportunities to those who were unemployed because of drought) and from nondrought investment strategies. The main items of this infrastructure are water resources, accumulated in dams or riverbeds through perennation [the process of making an object perennial] of some rivers; irrigation projects; transportation and communication facilities; urban facilities; and energy supply.

   Besides infrastructure, the economic development strategies of the Nordeste resulted at times in high rates of economic growth, with industrialization assuming an important role in the regional economy—and industry is much less vulnerable to droughts than dry-farming agriculture. A higher level of economic development also means more government revenues and thus more government capacity to finance relief actions during droughts.

   Infrastructure and economic development helped increase the Nordeste's resilience to droughts. But economic growth without an appropriate process of income distribution resulted in economic concentration and social exclusion, so that the stricken population remains poor and without proper means to defend themselves against the effects of the droughts.
2. In the last one hundred years, the population increased about seven times (from 6 million in 1890 to 40 million in 1990). Agricultural activity increased to the same

degree, or even more. But the natural vegetation has been devastated and the fauna has been sharply reduced. Urbanization has accelerated since the 1950s, and the demand for water has increased at a greater rate than the population as a result of high urban water demand.

Much more stress on the environment has resulted from the increased land use; this is a cause of dryland degradation and (eventually) desertification. From this viewpoint, the environment is now much less able to withstand drought events. That means that the strategy of looking for alternatives from the environment—hunting wild animals and digging roots for food—is now less successful. Even a light drought event becomes a serious problem, whereas in earlier times light events could be faced satisfactorily by the sparse population. This is probably one reason that drought figures of the past indicate that drought occurred less frequently than it does now. Indeed, the intense land use of the semiarid region, due to population growth and increased economic activities, as well as environmental degradation, reduces land carrying capacity and makes the region less resilient to climate variability.

CONCLUSION: SUSTAINABLE DEVELOPMENT AS A RESPONSE

In 1992, the International Conference on the Impacts of Climate Variations and Sustainable Development in Semi-arid Regions (ICID) recommended environmentally sustainable economic and social development as the ultimate goal to be pursued in semiarid regions (ICID, 1992). In the case of the Nordeste, this should be the strategy to make the region become permanently resilient to the problems of climate variability—present and future. The assumptions behind this statement are as follows:

1. For the population to become resilient, poverty must be eradicated and families must have access to a stable income that is enough to meet their basic needs and to provide for some level of savings. These conditions will enable them to face any economic crisis.
2. For the environment to become resilient, natural resources must be used in a conservative way to maintain and increase their productivity.
3. An equilibrium must be sought between the carrying capacity of the natural resources, the size of the population, and the amount and type of economic activities.

Sustainable development implies that economic development must continue, but its quality must change. This is a great challenge, because the concept of sustainable development, as stated by the Brundtland report (WCED, 1987), still lacks operational mechanisms—as well as financial mechanisms and appropriate technologies—that would allow developing and developed countries to adopt the idea. The Nordeste represents a large working field for scientists and policy makers at the national and international levels.

A sustainable development strategy requires—by definition—a long-term policy. This poses a challenge to policy makers used to making decisions on the basis of short-term political or resources constraints. A change of mentality is required.

The planning and implementation of a sustainable development strategy for the Nordeste should rely primarily on regional and national efforts. As in the case of other developing regions of the world, international cooperation is needed and its role is crucial.

ABOUT THE AUTHOR

Antonio Rocha Magalhães is an economist with the Esquel Brazil Foundation. He is a former state secretary of both Planning and Special Affairs in the state of Ceará, Brazil. He served as deputy superintendent of IPEA, the Brazilian Planning Institute of the federal government in Brasilia, and was also the special advisor to the Minister of Planning. Mr. Magalhães was the director of the International Conference on the Impacts of Climate Variations and Sustainable Development in Semi-Arid Regions, held in Fortaleza, Brazil, in 1992, in preparation for the United Nations Conference on Environmental Development (UNCED). He was one of the recipients of the Mitchell Prize on Sustainable Development (1991). He is a member of the Scientific Advisory Committee of the UNEP World Climate Impact Program (WCIP).

REFERENCES

Arraes, R.A.; and I. Castelar. 1989. The effect of drought on public finances in the state of Ceará. *In* A. R. Magalhães and E. Bezerra Neto. *Socioeconomic Impacts of Climatic Variations and Policy Responses in Brazil*. UNEP/SEPLAN-CE, Fortaleza, Ceará, Brazil (mimeo).

Braga, R. 1982. *História da Comissão Cientifica de Exploracão*. Fundação Guimarães Duque, Escola Superior de Agricultura de Mossoró Fortaleza. Coleção Mossoroense, Vol. CC.

Bitoun, J.; L. Guimarães Neto; and T. Barcelar. 1992. Amazonia and Northeast: The Brazilian tropics and the self-sustainable development. Paper presented at the International Conference on the Impacts of Climate Change and Sustainable Development in Semi-Arid Regions (ICID), Fortaleza, 27 January-1 February 1992.

DNOCS. 1985. *O DNOCS e o Novo Nordeste. Uma Perspectiva Histórica* (Anexo). NIEPCE-Núcleo Independente de Estudos e Pesquisas do Ceará, Fortaleza, Ceará, Brazil.

Duarte, R.R.S. 1989. The effect of power rationing in Ceará. *In* A. R. Magalhães and E. Bezerra Neto. *Socioeconomic Impacts of Climatic Variations and Policy Responses in Brazil.* UNEP/SEPLAN-CE, Fortaleza, Ceará, Brazil (mimeo).

Fraga, A.C., and A. T. Almeida. 1989. The effect of droughts on the economy of the state of Ceará. *In* A. R. Magalhães and E. Bezerra Neto. *Socioeconomic Impacts of Climatic Variations and Policy Responses in Brazil.* UNEP/SEPLAN-CE, Fortaleza, Ceará, Brazil (mimeo).

Gasques, J. G.; and A. R. Magalhães. 1987. Climate anomalies and their impacts in Brazil during the 1982-83 ENSO event. *In* M. R. Glantz and M. Krenz, eds. *The Societal Impacts Associated with the 1982-83 Worldwide Climate Anomalies.* NCAR/ESIG, Boulder, Colorado, USA; and UNEP, Nairobi.

Guerra, P.; and T. Guerra. 1909. *Seccas contra a Seca.* Fundação Guimarães Duque. Mossoró. Coleção Mossoroense, Vol. XXIX. Third edition, 1980.

ICID - International Conference on the Impacts of Climate Variations and Sustainable Development in Semi-arid Regions. 1992. *Declaration of Fortaleza.* Edited by Esquel Brasil Foundation.

IPCC - Intergovernmental Panel on Climate Change. 1990. *Potential Impacts of Climate Change.* WMO/UNEP, Geneva. (Report prepared for IPCC by Working Group II).

Luetzelburg, P. 1922-23. *Estudo Botânico do Nordeste*, Vols. I and II. Fundação Guimarães Duque and CNPq. Mossoró and Fortaleza. Coleção Mossoroense, Vol. CLXVI. Second edition, no date.

Magalhães, A. R. 1991. Regional impacts of climatic variations and policy responses: Droughts in Northeast of Brazil. *In* World Climate Programme-WCP. *Proceedings of the International Conference on Climatic Impacts on the Environment and Society (CIES)*; p. C13-18. University of Tsukuba, Ibaraki, Japan, 27 January-1 February 1991.

Magalhães, A. R.; and E. Bezerra Neto. 1989. *Socioeconomic Impacts of Climatic Variations and Policy Responses in Brazil.* UNEP/SEPLAN-CE, Fortaleza, Ceará, Brazil (mimeo).

Magalhães, A. R.; and Glantz, M. H., eds. 1992. *Socioeconomic Impacts of Climate Variations and Policy Responses in Brazil.* UNEP/SEPLAN-CE/ESQUEL, Brasília.

Magalhães, A. R.; and O. E. Rebouças. 1988. Introduction: Drought as a policy and planning issue in Northeast Brazil. *In* M. L. Parry, T. R. Carter, and N. T. Konijn, eds. *The Impact of Climatic Variations on Agriculture.* Vol. 2: *Assessments in Semi-Arid Regions*; pp. 271-380. UNEP/IIASA, Kluwer Academic Publishers, Dordrecht, Netherlands.

Magalhães, A. R.; J. R. A. Vale; A. B. Peixoto; and A. P. F. Ramos. 1989. Governmental strategies in response to climatic variations: Droughts in Northeast Brazil. *In* A. R. Magalhães and E. Bezerra Neto. *Socioeconomic Impacts of Climatic Variations and Policy Responses in Brazil.* UNEP/SEPLAN-CE, Fortaleza, Ceará, Brazil (mimeo).

Nobre, C. A.; and L. C. Molion. 1988. The climatology of droughts and drought prediction. *In* M. L. Parry, T. T. Carter, and N. T. Konijn. *The Impact of Climatic Variations on Agriculture*. Vol. 2: *Assessments in Semi-arid Regions*. UNEP/IIASA, Kluwer Academic Publishers, Dordrecht, Netherlands.

Pessoa, D. M. 1987. Drought in Northeast Brazil: Impact and government response. *In* D. A. Wilhite and W. E. Easterling, eds. *Planning for Drought: Toward a Reduction of Societal Vulnerability*; pp. 471-88. Westview Press, Boulder, Colorado, USA.

SUDENE. 1985. *A Problemática e a Política da Terra e Água do Nordeste*; Vol. 7. Recife (série Projeto Nordeste).

WCED - World Commission on the Environment and Development. 1987. *Our Common Future*. Brundtland Report. Oxford University Press, Oxford.

Chapter 11
Drought Policy in the Republic of South Africa

J. J. Bruwer

INTRODUCTION

Droughts are a regular feature of the African weather pattern. The incidence of drought (broadly defined as less than 70% of normal precipitation) in the western and northwestern part of the Republic of South Africa is about one year in three. The drought that commenced in 1978 affected, to varying degrees, about 75% of the area of the RSA (Baard, as cited by Wilhite, 1987). Average annual rainfall in the RSA exceeds 500 mm in only about 30% of the area. Thus, it is clear that considerable agricultural production takes place under arid and semiarid conditions and that large areas of the country are for much of the time under moisture stress.

Disaster droughts in Southern Africa should be regarded as recurring hazards and equal to many other agricultural constraints to be faced by the farmer. Minimizing the devastating effects of drought can only be done by proactive planning and regular evaluation of adapted strategies.

For one reason or another, sufficient effective countermeasures have not yet materialized, and with the onset of each drought, many land users are seemingly unable to cope with the adverse conditions. Today, a general consensus exists among agricultural leaders that most countermeasures are usually reactive in nature. Instead, there should be a strong proactive approach to coping with drought, and earnest attempts should be made to respond in a timely manner to the onset of the deteriorating conditions (Bruwer, 1989).

Furthermore, it should be fully appreciated that many of the impacts of drought are human-induced. Several examples illustrate this point. Aucamp (personal communication, 1989) states that over a specific assessment period, the tall grass veld and thorn veld areas of Natal were overstocked by 55% and 68%, respectively. Similarly, the thorn veld of the Eastern Cape was overstocked by 48%. It has also been estimated that the natural veld, for the country as a whole, is about 50-60% overstocked. Such overstocking has led to severe degradation of the natural veld.

Pasture research in the Orange Free State (OFS) over many years has clearly indicated that veld management can markedly influence the effects of drought. Fouche (1984) found that stocking rate largely influences the period, and probability, of feed shortages. Furthermore, veld condition is closely related to overall production and grazing capacity.

Other injudicious farming practices, such as the cultivation of marginal land (to such an extent that a special soil conversion scheme, by which croplands can be converted to pastures, has been introduced), have also contributed to the ravages of drought and to the greater vulnerability of the natural resources. Thus, humans have contributed significantly to environmental degradation and, in so doing, have greatly aggravated the effects of drought.

A combination of frequently occurring droughts and the inability of people to live and farm in harmony with nature, has led, inter alia, to the gradual impoverishment of many farmers and a concomitant deterioration of the natural resources. It is estimated that the total farming debt of South African farmers has escalated from ± R1,400 million in 1970 to approximately ± R14,000 million in 1989. It is predicted that if no deliberate action is taken to combat resource degradation, the Karoo biome could reach the Tugela river in Natal by the year 2050.

To combat such negative trends, the government has initiated various aid schemes over the years. For some, the efficiency of such schemes is questionable. Wilhite (1987) reports that in the 1984-85 fiscal year, the RSA government spent approximately R447 million in support of various drought relief programs. In 1987-89 and 1988-89, a record ± R1,300 million was allocated for drought and flood relief schemes by the state. This represents a substantial drain on the country's financial resources and reflects the serious consequences of natural disasters.

With this as background, the question may be posed whether such excessive amounts allocated to agricultural aid schemes are justified or effective in achieving the declared objectives. Concern for the continued degradation of natural agricultural resources in many parts of the country provides strong motivation for the government to continually review its policy and approach in offering financial and other aid to the agricultural sector.

DEFINITION, OCCURRENCE, AND ASSESSMENT OF DROUGHT

South Africa, like other countries, has suffered from the lack of a universally acceptable definition of *drought* (Wilhite and Glantz, 1985). Drought is recognized as a temporary feature of a region, the impact of which is felt over a period of time (i.e., it is a non-event). Three types are generally recognized (Thomas, 1965): *meteorological drought*, occurring when rainfall is abnormally low; *hydrological drought*, occurring when actual water supply is below the minimum for normal operations; and *agricultural drought*, occurring when soil moisture is depleted to the extent that crop (and pasture) yields are considerably reduced.

According to Schulze (1984), it is essential to distinguish between *drought* and *aridity*: both are characterized by a lack of water, but aridity carries the condition of a more or less permanent climatic condition while drought is a temporary condition.

Zucchini and Adamson (1984) hold the view that "drought occurs when there is less water available than is needed and not when there is less than expected."

In South Africa, drought is broadly defined as occurring at 70% of normal rainfall. It becomes a *disaster* or *severe* drought when two consecutive seasons experience 70% or less rainfall. A disaster drought implies that an area would qualify for government relief and experience a marked reduction in agricultural production concomitant with a high risk of stock or crop losses. A "normal" drought, as discussed in this paper, refers to temporary periods of moisture deficits of less than one year duration.

Ideally, drought should be studied in relation to its duration, intensity, spatial extent and time of occurrence within the annual cycle. Most attempts to describe the occurrence of local drought have been based on analyses of meteorological data.

The phenomenon of drought has been studied by many South African researchers over the past few decades and serious attempts, notably those by Tyson and Dyer (1975), have been made to identify apparent periodicities of rainfalls in the region. Most studies have centered on the apparent persistence of periods above or below average rainfall and the duration and frequency of occurrence of such periods. In contrast, Dent et al. (1987) studied the relative severity and areal extent of recent drought periods in the summer rainfall region. The rank order of the six driest periods on record (1925-84), for a single summer up to four consecutive summers, is presented in Table 1. These data suggest that for the period 1925-84, the drought of the two consecutive summers of the early 1980s was the most severe on record, according to the indices used. Other devastating droughts include those of the three consecutive summers ending in March 1933 and those of the late 1960s and early 1970s. The latter centered primarily on the central and western parts of the region, while that of the 1980s was located more to the east.

Continual assessment of drought conditions in South Africa is imperative. For this purpose, one of the indices used on a trial basis is the well-known Palmer Drought

Rank Order	Single Summer	2 Consecutive Summers	3 Consecutive Summers	4 Consecutive Summers
Driest	1982/83	1981/82–1982/83	1967/68–1969/70	1967/68–1970/71
2nd Driest	1969/70	1968/69–1969/70	1981/82–1983/84	1966/67–1969/70
3rd Driest	1967/68	1967/68–1968/69	1944/45–1946/47	1964/65–1967/68
4th Driest	1948/49	1945/46–1946/47	1963/64–1965/66	1965/66–1968/69
5th Driest	1972/73	1982/83–1983/84	1968/69–1970/71	1948/49–1951/52
6th Driest	1951/52	1964/65–1965/66	1930/31–1932/33	1969/70–1972/73

Table 1. Rank order of the six driest periods on record (1925-84) (after Dent et al., 1987).

Severity Index (PDSI), a single numerical value taking into account precipitation, potential evapotranspiration, soil moisture, and runoff (Palmer, 1965). One of its main limitations is that it is not generally indicative of short-term (several weeks) changes in moisture status affecting crops and farming operations. Furthermore, the lack of reliable meteorological data in the arid areas of the country resulted in the index being unsuitable for making decisions regarding drought relief schemes.

More recently, Erasmus (1988) has determined rainfall deciles for monthly and yearly records, involving more than 4,000 stations, with at least 20 years' data, as a means of monitoring drought. A data bank of more than 140 digital maps of South Africa was created. These maps, displayed on an image processor, are used to identify the presence of drought conditions and to monitor their development, intensity, and spatial extent.

The climate and rainfall of the semiarid subcontinent of Southern Africa has been highly variable for millennia and will continue to be so in the future. All studies clearly indicate that drought of varying severity is a regular feature over the entire region. *All future planning must be predicated on the assumption that it is a land of drought rather than a land of plentiful rain* (Tyson, 1988). This means that greater attention must be given to drought planning and the implementation of measures to counter the impact of moisture deficiencies.

DROUGHT POLICY

Droughts of disaster magnitude must be accepted as a feature of the environment. The justification for drought assistance schemes is outlined in the following paragraphs, with particular reference to the Republic of South Africa.

In any country, the state has a moral obligation and responsibility to assist individual citizens and communities during times of hardship resulting from natural disasters. In the case of disaster droughts, the state undoubtedly has justification for intervening to prevent communities and infrastructure from disintegrating. In this way, disruption of agricultural and socioeconomic development in rural areas can be avoided. To fulfill this obligation, the state should act discreetly in distinguishing between disaster situations created as a result of human actions and those that are purely the result of natural phenomena. Any state assistance should, therefore, be planned and directed on sound scientific and economic principles, including the accepted optimal agricultural resource utilization approach. A well-conceived drought policy is essential and should express the *method, philosophy*, and *criteria* on which the government, through a responsible State Department, can base its actions to assist the severely affected communities.

Since regular occurrence of disaster droughts is a reality, it is essential that farmers *plan* for such droughts. Such planning includes *long-term* and *short-term actions*. The long-term actions include the implementation of a national grazing strategy and enforcement of acts such as the Conservation of Agricultural Resources Act of 1983 and the Subdivision of Agricultural Land Act of 1970, which are aimed at reducing the negative impact of a disaster drought by maintaining and improving pastures. The short-term actions imply the implementation of disaster drought assistance schemes for stock farmers.

Preferably, farmers should act *proactively*, with sound management methods to handle disaster droughts, instead of *reactively*. During disaster droughts, the government ideally would grant assistance to farmers who had conserved agricultural resources proactively but who, as a result of factors beyond their control, were adversely affected by a disaster drought.

THE AIMS OF A DISASTER DROUGHT ASSISTANCE SCHEME

A disaster drought scheme should primarily be aimed at protecting natural resources, and therefore it should constantly encourage farmers to apply the three basic requirements of an optimum or sustainable resource utilization system.

1. Agricultural production must be in harmony with natural environmental factors.
2. Agricultural production may not be achieved at the expense of the agricultural resources.
3. Agricultural production must be on an economic basis.

A disaster drought scheme should also support farmers when a drought has assumed such disaster proportions in a specific area that the state must reasonably assist farmers in bridging the disaster drought financially, contribute to the maintenance of a nucleus breeding herd/flock within the limits of the production capabilities of natural agricultural resources, and grant financial and other assistance in such a way that correct management of agricultural resources is promoted.

PRINCIPLES ON WHICH THE DISASTER DROUGHT ASSISTANCE SCHEME IS BASED

The farmer must accept responsibility for bridging seasonal droughts that are inherent to the environment and do not assume disaster proportions. Measures introduced against these droughts must in the future *be regarded as an integral part of the farming management program.*

Stock farmers will also be expected to be self-sufficient to a much greater extent in the initial phases of a disaster drought by implementing sound farming management practices—in other words, they must take proactive action.

DEFINITION OF A DISASTER DROUGHT

A disaster drought is defined as a farming condition in which production and availability of natural and cultivated pastures, fodder production, and water supplies in a specific area have deteriorated to the extent that natural agricultural resources and livestock production are seriously affected and livestock mortalities are expected to occur if livestock numbers are not further drastically reduced and survival rations supplied to the remaining animals. It is the result of unfavorable climatic conditions, of which subnormal rainfall over one or more rainy seasons is the major causal factor.

A disaster drought condition should not be attributed to the exceeding of the long-term grazing capacity and/or carrying capacity of an area or a farm. Rather, a timely, gradual stock reduction should occur as conditions deteriorate and no further feed reserves are available.

The introduction and termination of disaster drought assistance in an area, or portion thereof, must be based on specific and objective criteria. An evaluation scale or point system has been developed in South Africa in which the intensity of a drought can be determined relatively objectively (Roux, 1991). District drought committees (DDC) now evaluate drought intensity according to a numerical rating scale. This rating is then used as a basis for the introduction and termination of disaster drought areas. The points chart (evaluations scale) takes into consideration the unique characteristics of broad homogeneous farming areas. Factors evaluated for each farm include climatic conditions, veld status, pasture and crop status, livestock status, and water availability. The condition of the natural agricultural resources (especially in respect to pastures and feed supply) on *conservation reference farms* is used as a benchmark against which farms being evaluated for relief are judged. These conservation reference farms should be well distributed over the specific disaster drought area. The DDCs should be responsible primarily for identifying these conservation reference farms, and they must consult with conservation committees in the area. The conservation committees may also be directly involved in the identification of these farms. The conservation reference farms are selected on a "permanent" basis, but they are supplemented with farms that are "temporarily" used for assessing a specific application if "permanent" conservation reference farms have not been identified in the area. The names of conservation reference farms must be noted in a register, which is kept by the DDC. The implication is that in the future an area will not be listed as a disaster drought area until conditions on these conservation reference farms justify disaster drought assistance.

INSTRUMENTS TO IMPLEMENT DROUGHT POLICY

To implement a specific drought scheme, the following administrative instruments and measures are needed:

- A national drought committee (NDC) should advise the Minister of Agriculture on all matters concerning drought assistance and should scrutinize applications for the proclamation and termination of drought-stricken areas (districts or parts thereof).
- At the local farm level, DDCs are needed to consider all applications for the proclamation or termination of disaster drought areas according to specified guidelines; it must submit its recommendations to the NDC.
- The grazing capacity of all natural pastures should be determined and zones with similar *grazing capacities* demarcated and used as a basis for calculating incentives and so forth. (*Grazing capacity* refers to the number of hectares per livestock unit that can be kept and maintained only on the natural veld/grazing.)

ASSISTANCE MEASURES UNDER THE DISASTER DROUGHT ASSISTANCE SCHEME

Assistance measures may consist of the following four components:

- A *rebate* on the transport of stock feed.
- A *state contribution* to the maintenance of a nucleus herd.
- An *incentive* on stock reduction.
- A *lease scheme*.

Payments for the state contribution, incentive, and lease scheme are made per converted large stock unit (LSU) and are subject to the availability of state funds. The Minister of Agriculture will announce the valid tariffs from time to time.

The components of the assistance measures are discussed below.

Rebate

A 75% rebate on transport costs will be granted to participants for the transport of approved stock feeds and licks by rail or authorized and licensed haulers. The rebate is paid on the transport costs from the place where the participant buys the fodder to his farm(s). *The haulers will not give any discount.* Invoices/proof of the transport costs of the stock feeds and licks must accompany the monthly claim to qualify for the rebate—

in other words, participants must pay the transport costs themselves and subsequently claim the rebate.

The DDC must confirm that the tariffs of authorized and licensed haulers are within reasonable limits. No rebate is granted on the transport of livestock.

State Contribution to Maintenance of Nucleus Herd/Flock

A state contribution to feeding costs is linked to the maintenance of a *nucleus breeding herd/flock*. (A *nucleus herd/flock* is defined as one-third of the permissible number of livestock with which a participant may farm according to the long-term grazing capacity and/or carrying capacity of his farm[s].)

The state contribution is made to either a *maximum limit* of one-third of the permissible number of livestock with which a participant may farm or 125 LSU, whichever is the smallest. This means that the state contribution to a participant who may keep 400 LSU will be limited to 125 LSU.

Similarly, a participant who may keep 300 LSU will receive a state contribution of a maximum of 100 LSU. The contribution concerned, however, will only be allocated according to the actual number of converted LSU on the livestock inventory of the farm for the month for which the claim is submitted. If, therefore, a participant who may keep 300 LSU had only 75 LSU on the farm for a specific month, the state contribution will be made for 75 LSU.

The amount paid per LSU is calculated according to a specific formula that includes market-related feed prices. The amount will be adjusted according to changes in feed prices and the budget allocation.

Incentive

An incentive amount is paid for the reduction of livestock. In calculating the amount according to a specific formula, the long-term grazing enterprise is used as the basis. A participant receives a monthly amount for each LSU that is less than that permitted according to the grazing capacity or carrying capacity.

To qualify for participation, at least one-third of the permissible load must be reduced. Reductions of more than one-third, or even total reduction, is permissible. In fact, it is desirable to make reductions of more than one-third as disaster conditions worsen and the natural resources decline.

The incentive amount paid per reduced LSU is linked to the average net farming income (NFI) yielded by small stock (sheep and goats) and large stock (cattle), respectively. This amount will also fluctuate over time and be adapted as the NFI changes and according to the availability of funds.

The *incentive amount per reduced LSU* will, in accordance with the amount determined for small stock and large stock, be *the same* from the withdrawal of the first LSU until one-third remains.

When a participant reduces livestock to below one-third of his unit's grazing capacity or carrying capacity (in other words, when a reduction of more than 66.7% is applied), he still receives incentive payments on the number of LSU reduced below that quantity. Because the participant then no longer complies with the objective of the maintenance of a nucleus herd/flock, the incentive is reduced. The reduced incentive only applies to the portion reduced below the last one-third. For a withdrawal from the basic number down to two-thirds (66.7%), the full or ordinary incentive is paid.

Lease Scheme

A state monetary contribution per LSU per month will be paid for the lease of grazing elsewhere. In a proclaimed disaster drought area, two-thirds of the stocking rate that the participant may keep on his farm (according to the grazing capacity or carrying capacity of his farm) will qualify for this lease scheme.

The conditions of the lease scheme are as follows:

- The properties on which the grazing or pastures are leased must be situated outside a proclaimed disaster drought area at the beginning of the term of lease.
- A farmer may not participate in the lease scheme and the disaster drought assistance scheme simultaneously and receive benefits from both. This means that a farmer must choose between the lease scheme and the disaster drought scheme. Should the leased farm become so dry that it is also proclaimed as a disaster drought area, participation in the lease scheme may be suspended and the participant may take part in the ordinary disaster drought assistance scheme.
- For the period during which he participates in the lease scheme, a farmer must withdraw all his property in the disaster drought area from grazing. Livestock may only be returned to the disaster drought property two months after the property is no longer in a disaster drought area.
- The property leased must be able to accommodate the participant's livestock (two-thirds of his property's carrying capacity in the disaster drought area). The grazing capacity/carrying capacity of the leased property may not be exceeded. A livestock register must be submitted monthly.

FURTHER CONDITIONS OF THE DISASTER DROUGHT ASSISTANCE SCHEME

- To ensure that state assistance is given only to farmers who implement conservation principles, it is a proviso that stock farmers who plan to participate in future disaster drought assistance schemes must register as prospective participants from a certain date for those farms that will be involved.
- Participants will at times be allowed to exceed the long-term grazing capacity/carrying capacity, but within the norms of a scientifically based grazing system acceptable to the Department of Agriculture. However, the recorded *average stocking rate over the preceding 12 months before participation may not exceed the grazing capacity/carry capacity.*
- Should a prospective participant repeatedly exceed the average long-term grazing capacity/carrying capacity over the 12 months, *the participant will be disqualified from participating in the scheme for 12 months after its introduction.* Such a person will therefore only be able to start participating after a year, provided that he complies with the conditions in the meantime.
- *Failure to keep the prescribed registers up to date also disqualifies a prospective participant for 12 months* from the introduction of the scheme until the registers have been kept up to date for at least 12 consecutive months.
- Farmers who have just started farming will have to register immediately if they wish to become prospective participants. Should a disaster drought develop within the first 12 months, they will be able to qualify for participation on the grounds of the records already submitted, even if these are for less than 12 months. If new farmers have already farmed for more than 12 months, they will also be subject to the proviso that records for 12 months must already have been submitted.
- All farmers—full-time or part-time—in a proclaimed area may apply, but assistance is granted only to farmers whose units, in the opinion of the DDC, are truly established and operated as farming enterprises. Drought assistance will NOT be granted to a soil user who can accommodate fewer than 30 converted LSU according to the long-term grazing capacity or carrying capacity of the unit.
- Apart from individual farmers, farming companies, partnerships, closed corporations, and trusts will also qualify. If a participant belongs to more than one closed corporation and/or trust, the maximum state contribution to the maintenance of a nucleus herd/flock for which such a person qualifies will be limited to 125 LSU. Participants, regardless of the number of farming enterprises that they operate or in which they may have an interest, will only be able to *submit one application* for the maintenance of a nucleus herd/flock—not for each enterprise separately. The

rebate and the incentive are not restricted and apply to all units participating in the disaster drought assistance scheme that maintain the permissible stocking rate and concomitant stock reduction applicable to it.

- If more than one person on one farm and/or farming enterprise operates a recognized, proven, separate, and independent enterprise, each person qualifies for participation in the scheme. The DDC will have to control and confirm that each person who claims separate assistance does, in fact, operate a separate, independent enterprise.
- Tenants of farms in the disaster drought areas will have to submit a valid lease contract together with the first applications in respect of property leased and which are entered under the disaster drought assistance scheme. The DDC must ensure that the contract is valid. The term of the lease must be valid for at least one year from the date of participation.
- The Department of Agriculture reserves the right to decide whether or not an applicant qualifies for assistance, on the understanding that such a decision may be appealed to the Minister of Agriculture.
- The *incentive* on stock reduction is paid from the start of participation and is adjusted monthly as the livestock is reduced. To qualify for the incentive, the livestock numbers at the start of participation may not exceed the long-term grazing capacity/carrying capacity. If they do, no incentive will be paid until the stocking rate is less than the grazing capacity/carrying capacity. The stocking rate must be reduced by at least one-third *within four months* to qualify for further participation. If the stocking rate is not reduced by at least one-third within four months, penalties will apply.
- *The rebate and state contribution* to the maintenance of a nucleus herd/flock will only be paid to an applicant after his stocking rate has already been reduced by at least one-third of his property's official grazing capacity/carrying capacity. No rebate or state contribution will be paid before that time, nor will it be retroactive. For example, if an applicant's livestock registers show that he had reduced his stocking rate by one-third on April 17, the benefits come into effect from May 1.
- It is desirable that the livestock removed from herds/flocks be sold. This is not a proviso, but the livestock must be removed from the farm or farms concerned. Where the grazing or pastures are leased, the property concerned must, at the start of the lease period, be situated outside a proclaimed disaster drought area. The long-term grazing capacity/carrying capacity of the property may not be exceeded. A livestock register of the leased farms must also be submitted monthly.
- No invoices of feed purchases are required for the state contribution to the maintenance of a nucleus herd/flock. However, participants seeking payments for feed for livestock that have been withdrawn from herds and placed in feedlots must submit invoices of feed purchases to qualify.

- Farmers must keep livestock inventory forms up to date monthly and attach them to the application for state contribution/incentive/rebate.
- The present enumeration of Act 43 of 1983 is applied. In other words, an unweaned lamb, calf, and foal, and an ostrich chick, are included in the calculation of the stocking rate.

TERMINATION OF THE DISASTER DROUGHT ASSISTANCE SCHEME AND AFTER-CARE ACTION

The formal termination of the disaster drought assistance scheme must occur in a timely manner according to the guidelines for the introduction and termination of disaster drought areas, as explained in the relevant manual. To offer the grazing and perennial cultivated pastures a chance to recover after a disaster drought, a specific after-care action must be taken. Participants must, when applying for participation, give a written statement as to how this action will be carried out (implemented), and this must incorporate the following practices: For two months after termination of the scheme, a participant must maintain the same (or lower) stocking rate as the one that applied at the time of termination. This implies that, apart from rams, bulls, and stallions for breeding purposes, livestock may not be increased by purchases. Should the termination occur during the lambing, calving, and foaling season, increase via natural breeding is permissible. The stocking rate may not, however, exceed two-thirds of the unit's grazing capacity; in other words, a one-third reduction must be maintained for at least the full two months. If the animals are kept on grazing and/or cultivated pastures, no form of remuneration will be paid to participants after the date of termination. Should a participant, after rain occurred that led to the termination of the disaster drought scheme, continue to keep all the livestock in feedlots to give the grazing a better chance of recovery, the rebates, state contributions, and incentives will still be paid to such a participant for another two months. Farmers wishing to follow these practices must, within one month after the date of the termination of the disaster drought assistance scheme for stock farmers has been announced, make written application for further participation to the DDC, and implementation of the action must be controlled by the DDC. Farmers who implement this specific after-care action have the option to implement voluntary stock withdrawal for an additional two months over and above the previous two months. In this case, at least a one-third reduction must be maintained, but the livestock may be kept on the grazing or cultivated pastures. During these additional two months, only an incentive compensation for the reduced livestock numbers will be paid. Application must be made to the DDC, which can give the necessary approval for the sake of veld and grazing recovery. The after-care action to which remuneration is attached could therefore continue for four months after termination of the scheme.

CONCLUSION

A disaster drought aid scheme that is based on a well-conceived drought policy will contribute significantly to the following:

- Prevention of the continuation of natural resource degradation.
- Maintenance of agricultural production that will help ensure that no major food shortages occur.
- Maintenance of rural communities and infrastructure, thus counteracting the possible depopulation of rural areas.
- Increase in employment.
- Reduction of political pressure on the government.
- Creation of a cooperative climate between agricultural interests and the state department(s) concerned, resulting in the mutual acceptance of the responsibility to cope with disaster drought situations.
- Stimulation of agricultural risk management rather than crisis management.
- Efficient use of state relief funds for communities.

ABOUT THE AUTHOR

J. J. Bruwer is the former chief director of Farming Development, Department of Agriculture and Water Supply, Pretoria, Republic of South Africa. He also served as the chairman of the National Drought Committee. Under his leadership, many of the recent changes in the government's drought policy occurred.

REFERENCES CITED

Bruwer, J. J. 1989. Drought policy in the Republic of South Africa. *Proceedings of the SARCCUS Workshop on Drought*. Alpha Training Center, South Africa.

Dent, M. C.; R. E. Schulze; H. M. M. Wills; and S. D. Lynch. 1987. Spatial and temporal analysis of the recent drought in the summer rainfall region of Southern Africa. *Water SA* 13:37-42.

Erasmus, J. F. 1988. *Drought Monitoring: Using Rainfall Deciles as a Drought Index*. Report GB/A/88/6. Soil and Irrigation Research Institute, Pretoria, South Africa.

Fouche, J. J. 1984. Ondersoek in die gebruik van die PUTU 11 simulasiemodel en Palmerindeks vir die karakterisering van droogtetoestande. M.S.-verhandeling U.O.V.S.

Palmer, W. C. 1965. *Meteorological Drought*. Research Paper No. 45. U.S. Weather Bureau, Washington, D.C.

Schulze, R. E. 1984. Hydrological simulation as a tool for agricultural drought assessment. *Water SA* 10:55-62.

Thomas, H. E. 1965. Reality of drought is always with us. *Natural History* 74:50-62.

Tyson, P. D.; and F. G. J. Dyer. 1975. Mean fluctuations of precipitation in the summer rainfall region of South Africa. *South Africa Geographical Journal* 57:124.

Wilhite, D. A.; and M. H. Glantz. 1985. Understanding the drought phenomenon: The role of definitions. *Water International* 10:111-20.

Wilhite, D. A. 1987. The role of government in planning for drought: Where do we go from here? *In* D. A. Wilhite and W. E. Easterling, eds. *Planning for Drought: Toward a Reduction of Societal Vulnerability*; pp.425-44. Westview Press, Boulder, Colorado.

Zucchini, W.; and P. T. Adamson. 1984. *The Occurrence and Severity of Droughts in South Africa*. Water Resources Commission Report 91/1/84. Pretoria, South Africa.

Chapter 12
Drought in Australia: Prediction, Monitoring, Management, and Policy

David White, Drew Collins, and Mark Howden

INTRODUCTION

Drought in Australia affects not only the rural sector but also the economy in general (Campbell et al., 1983). The Australian government, in collaboration with the states, has therefore set about developing and implementing policies to help farmers and other natural resource managers improve their ability to prepare for and manage drought.

At the farm level, it is essential to develop and implement sustainable systems that minimize the impact of drought on the soils and vegetation and ensure that livestock are humanely cared for or disposed of. The well-being of farming families will also be enhanced through better farm, financial, and risk management. The highly variable climate emphasizes the importance of improving drought forecasts.

A range of computer- and satellite-based technologies is now available to improve the self-reliance of land managers. Computer-based tools, including Decision Support Systems, should help farmers to cope with climatic variability, so that the physical, biological, and financial consequences of different management strategies are better understood. Although the threat of drought cannot be removed, its impact on the community, and on soils, vegetation, and livestock, may be reduced.

Simulation models may also be used to estimate changes in soil moisture and vegetative cover. Remote sensing by satellite can be used to monitor the soils and vegetation and indicate when plant productivity is starting to be restricted by lack of moisture.

In this chapter we outline features of the Australian climate and its impact on agriculture. The economic significance of drought, reasons why past government assistance led to inequities and poor management practices, and essential characteristics of the new national drought policy are discussed. Emphasis is placed on how government policies, backed up by technologies to aid drought prediction, monitoring, and management, are expected to lead to more self-reliant management at the farm level, and the development of agricultural systems that are physically, biologically, and financially sustainable.

FEATURES OF THE AUSTRALIAN CLIMATE

Australia is an island continent with ancient, fragile soils and a predominantly arid climate. Rainfall above 600 mm per annum is confined to the northern, eastern, and southeastern coastal regions and the southwestern tip (Figure 1). The south is characterized by a Mediterranean-type climate with cold, wet winters and hot, dry summers. Annual pastures in the south usually germinate between March and May in response to autumn rains, growth being most active in spring before senescence in October-November. The north experiences a monsoonal climate, most of the rainfall occurring in late summer-autumn (January-May).

There is substantial variability in the rainfall between years, this variability increasing in general with distance inland (Figure 2). With climate variability there is inevitably drought. Australia therefore has the reputation of being a land of droughts and flooding rains.

Agriculture in Australia is predominantly extensive. In the limited area where high rainfall occurs, above 800 mm a year, agriculture in southern and eastern Australia is

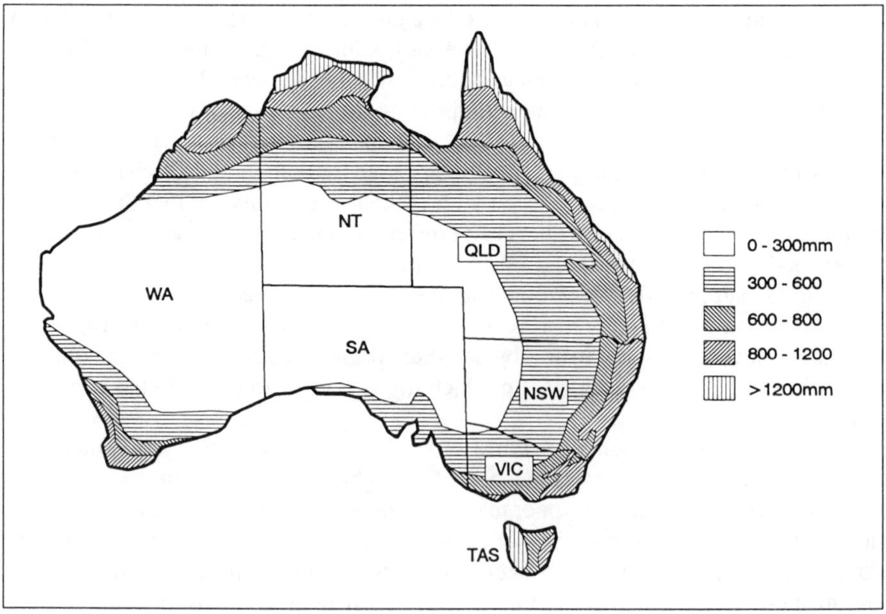

Figure 1. Map of mean annual rainfall isohyets for Australia. Source: Bureau of Meteorology.

characterized by dairy and beef cattle, horticulture, and prime lamb production. Most wheat production, in association with sheep, is located in the southeast and southwest, between the 300 and 600 mm annual rainfall isohyets, though wheat production farther north moves to higher rainfall areas. The balance of Australia's wool and beef production takes place in the pastoral zone, most of which has considerably less than 600 mm rainfall a year. A large part of the center of Australia is desert or very arid rangelands that in most years are able to support relatively few grazing animals.

Most of Australia's rural wealth is generated in the high rainfall and wheat-sheep areas. The economic significance of drought on the economy is therefore greater when these areas are affected.

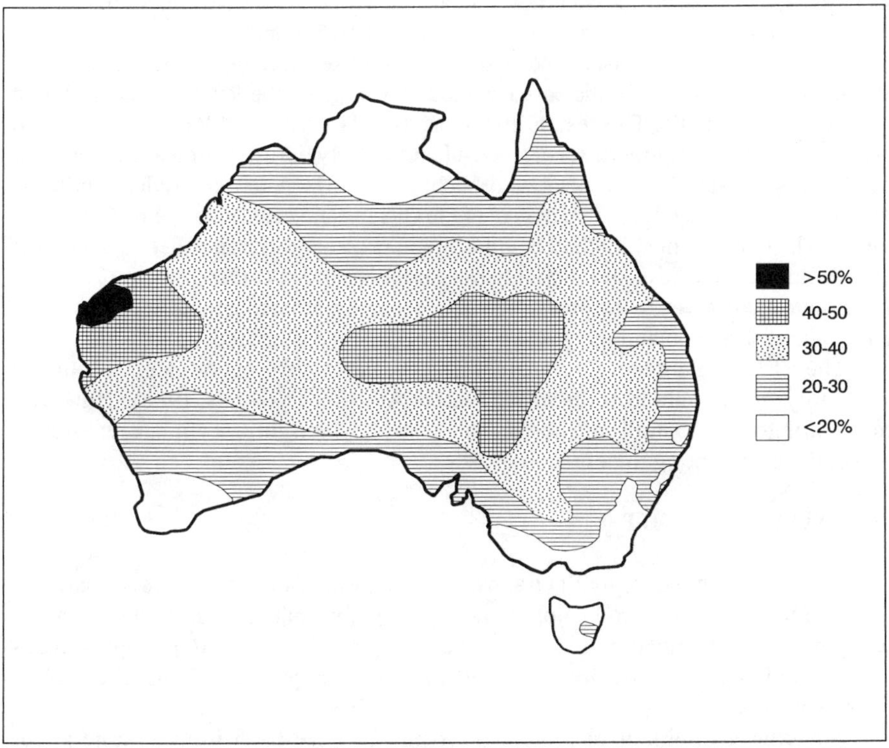

Figure 2. Map of rainfall variability (coefficient of variation of annual rainfall, %) for Australia (Barkley, 1931; cited in Gentilli, 1971).

CAUSES OF DROUGHT

The physical causes of drought have their origins in the natural fluctuations of the global climate system, this being influenced by interactions in time and space between the atmosphere, oceans, ice masses, and biosphere. At many locations, rain appears to occur at random, droughts and floods being stochastic events. However, the more knowledge we have of how the various elements of climate interact, the better we are able to understand the causes of the larger-scale droughts (Hunt, 1991).

The most widely known recurring climatic anomaly related to drought is the El Niño phenomenon. El Niño, literally "the Boy Child," is the name initially given to the warm current that appears periodically off the coast of Ecuador and Peru around Christmas. This coastal current is now recognized as part of a global climatic event that happens about every 3-6 years and typically lasts 12-15 months.

El Niño events are associated with changes in sea-surface temperatures in the Pacific and a slackening of the easterly trade winds across the Pacific between South America and Australia. This results in increased rainfall in the mid-Pacific and reduced rainfall in spring in southern Australia, and particularly during spring and summer in northern Australia (e.g., McBride and Nicholls, 1983). A marked reduction in the number of tropical cyclones that pass over the Queensland coast also occurs (Hastings, 1990). The concurrent shifts in the atmospheric pressure difference between Darwin and Tahiti form the basis of the Southern Oscillation Index (SOI) (Troup, 1965). La Niña is the name given to the phenomenon in which the atmospheric differences are in the opposite direction and increased rainfall occurs in eastern Australia.

The El Niño/Southern Oscillation (ENSO), hence, is a major cause of climate variability and drought in northern and eastern Australia (Nicholls, 1988). However, many droughts in Australia are due to other causes, which current climate research in Australia is attempting to define.

DEFINING DROUGHT

Drought may be simply defined as below-average rainfall that restricts typical plant growth for agricultural production. However, it is difficult to identify and measure drought by the quantity of rain alone, as the timing and frequency of precipitation, as well as soil type, topography, and land management practices, will affect plant responsiveness.

Nevertheless, rainfall measurements typically have been used as criteria for defining drought. Specifically, annual rainfall in the first decile provides the basis of the Australian Bureau of Meteorology's *Drought Review* and is the basis for the maps in Figure 3. Considerable spatial variation exists in the incidence and extent of

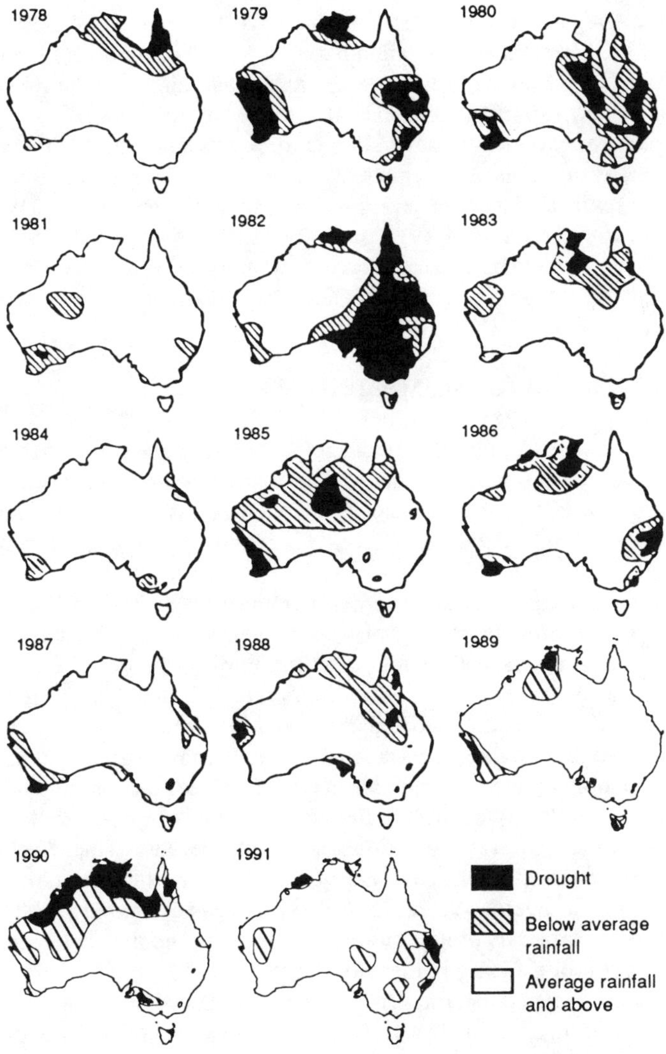

Figure 3. Major drought-affected areas in Australia, 1978-91; updated from Reynolds et al., 1983. Source: Australian Bureau of Meteorology.

droughts, not only in latitude between the temperate and tropical zones, but also between eastern and western Australia (Heathcote, 1991).

At an aggregated level, several studies have lent support to this definition of drought, by showing a high correlation between published reports of the incidence of drought based on production criteria and with annual rainfall in the first decile (e.g., Gibbs and Maher, 1967; Hutchinson et al., 1992).

Australia has experienced several major droughts since rainfall records began in the 1860s, many droughts of lesser extent and severity having also caused appreciable economic losses in large areas of several states. Major droughts have occurred in 1864-66, 1880-86, 1888, 1897, 1914-1915, 1927-29, 1940, 1944, 1961, 1965-67, and 1982-83. Historically, it could be expected that in only about one year in five would Australia be completely free of areas with rainfall that is very much below the average (Reynolds et al., 1983).

ECONOMIC SIGNIFICANCE OF DROUGHT

Drought disrupts routine farm management practices designed to accommodate normal periods of rainfall deficiency. The disruption to livestock and cropping operations can lead to serious financial difficulties associated with reduced farm incomes and increased indebtedness. If this disruption is prolonged, it can threaten farm viability.

The major drought across eastern Australia in 1982-83 has been extensively studied. During the drought, the area of crops harvested was significantly reduced, and more grain than usual was retained for stock feed (Purtill et al., 1983). Failed crops were either cut for hay or grazed by stock. Most farmers decided to retain normal stock levels in anticipation of autumn rains and higher prices. Most drought-affected properties had to purchase feed despite the use of failed crops and roadside grazing. Often, this was made possible only by government assistance for fodder purchase and transport.

Purtill et al. (1983) estimated that farm incomes for Australia's broadacre properties were reduced an average of 45% in 1982-83 because of the effects of drought. However, this average masks regional variations. For example, farm incomes in Victoria fell from an average $A22,000 in 1981-82 to an estimated $A740 in 1982-83. In addition, debt carried by drought-affected farms between June and November 1982 increased about four times more than nonaffected farms.

It has been estimated that the 1982-83 drought led to a fall of 18% (or $A1520m) in gross rural output, resulting in a 1.1% decline in Australia's national output (Gross Domestic Product). The value of rural exports was estimated to have fallen by $A500m because of the drought, representing some 3% of total Australian exports (Campbell et al., 1983). Total employment was estimated to have fallen by 2% because of the drought.

In addition, drought has been associated with episodic land degradation events that typically have involved major losses of topsoil.

HISTORICAL OVERVIEW OF DROUGHT POLICY

Producer and government responses to drought have focused on drought as a national disaster. This has been fostered by the use of drought declaration procedures by state governments to trigger various assistance measures. Drought assistance provided by state governments has generally consisted of concessional loans for carry-on purposes; freight subsidies for the transportation of fodder, stock, and water; and financial aid for certain purposes.

Before 1 July 1989, Australian (federal) government drought assistance was also provided under National Disaster Relief Arrangements (NDRA). Under the NDRA, which operated for 30 years, the Australian government provided funding to the states after a threshold of expenditures had been exceeded. In addition, the Australian government has introduced specific measures in the past, such as the fodder purchase and interest rate subsidy schemes in 1982-83. It has been estimated that drought assistance of nearly $A300 million was provided by governments in 1982-83, of which some $A119 million was provided by the Australian government.

The NDRA scheme was probably superior to that operating in many other countries in that it did set out the Australian and state government guidelines before a natural disaster occurred. However, the incorporation and operation of drought assistance within NDRA have been criticized for several reasons. First, drought declarations were subjective and have sometimes been excessive from a historical perspective. Second, abuse of fodder and freight subsidies was evident. Also, these subsidies favored larger livestock properties, and the benefits were largely captured by suppliers through higher prices. The Bureau of Agricultural Economics estimated that only 51% of the $A157m fodder subsidy in the 1982-83 drought was actually received by the target group, and when the effect of the subsidy on intensive livestock producers was considered, a net social loss of $A35m was determined (Stoneham et al., 1986). Third, input-based subsidies provided poorly targeted assistance for both recipients and their needs. Fourth, and perhaps most important, these subsidies served to distort the relative costs and returns of alternative practices. In so doing, they did not foster either a philosophy of self-reliance or acceptance of drought as a normal feature of Australian agriculture. They also encouraged farmers to retain higher than desirable stocking rates as drought developed, leading to additional land degradation.

In summary, the operating environment of farmers was one in which assistance was expected, and farm operations were planned on the basis that governments would provide a "safety net" against climatic variability.

REVIEW OF DROUGHT POLICY

Agriculture in Australia is largely unsubsidized and, with the rest of the economy, is being further deregulated. Nevertheless, government recognizes that it still has a role in ensuring that the operating environment within which farmers work is conducive to sound business decision making. In particular, appropriate macroeconomic and microeconomic policies, income taxation, and industry policies need to be in place (Anonymous, 1990). Drought policy should not be developed in isolation from other policies that affect farmers. A holistic or systems approach to wider policy formulation is therefore required, so that any drought policy should be integrated with other policy areas such as ecologically sustainable development (ESD), income taxation, social security, and education and training.

In April 1989, the Australian government announced the establishment of an independent task force to undertake a comprehensive review of national drought policies. The government also announced that from July 1989 it would no longer provide drought relief under the NDRA. From this time until August 1992, drought assistance at the national level was provided under interim arrangements through the Rural Adjustment Scheme (RAS).

Under the interim arrangements, the RAS had three components. Part A provided concessional finance as an interest subsidy, loans, or grants to farmers who were assessed as having prospects of long-term profitability, but who were experiencing short-term difficulties. Part B provided carry-on assistance in the form of a 50% interest subsidy to farmers in industries or regions experiencing a short-term downturn beyond their control. Part C provided assistance to farmers assessed as nonviable in the long term to enable them to make an orderly exit from farming. In 1991, Part C was expanded to provide income assistance to farm households.

A NEW NATIONAL DROUGHT POLICY

The Drought Policy Review Task Force, in a report released in May 1990, endorsed the Australian government's objective of promoting a self-reliant risk management approach to drought and providing RAS-type assistance to some producers in severe drought.

It was recommended that drought should no longer be regarded as a natural disaster but rather as an integral part of a highly variable climate. It was suggested that government policy should be aimed at helping farmers to cope with this variability, particularly through assisting them to improve their decision-making frameworks. It was therefore recommended, and accepted by Federal Cabinet, that drought policy in the future should be based on the following three objectives: (1) to encourage primary producers and other segments of rural Australia to adopt self-reliant approaches in managing for climatic variation, (2) to facilitate the maintenance and protection of Australia's agricultural and resource base during periods of increasing climatic stress, and (3) to facilitate the early recovery of agricultural and rural industries to levels consistent with long-term sustainable production.

These recommendations recognize that assistance that attempts to protect farmers and possibly support unsustainable farming systems in the face of drought is misplaced. Rather, management practices that account for and cope with drought and which follow the principles of sustainable agriculture are needed. The new National Drought Policy is specifically concerned with the risks that drought poses for agricultural enterprises. It has an ultimate goal of productivity growth, better allocation of resources, and producer self-reliance, but it recognizes the role of government in assisting farmers to achieve this objective.

Achieving agreement on the content of the new National Drought Policy has involved extensive negotiations between the Australian government and the states. Policy measures introduced in this policy (Crean, 1992) include:

New Rural Adjustment Scheme

Ongoing incentives through interest subsidies on commercial finance or grants are intended to enhance profitability through farm improvement and development, training, farm appraisal and planning services, and (in special circumstances) carry-on financial support as part of a program of productivity improvement. In times of exceptional downturn, interest subsidies above the level of 50% of the interest rate on commercial loans or existing debt may be provided. In addition, farmers who are in extreme financial hardship and who cannot obtain commercial finance for carry-on purposes will be able to access farm household support.

Financial Management

Current income smoothing mechanisms, such as income averaging, will be complemented by an improved Income Equalisation Scheme and Farm Management Bond provision, which will encourage the development of a financially self-sufficient farm sector. Rural taxation provisions are to be examined in 1993-94 to ensure that these promote the adoption of sustainable practices, encourage expenditure on drought preparedness, and reduce the need for government assistance in times of drought.

Whole-Farm Planning

The concept of whole-farm planning is to develop an overall property strategy so that the farm can withstand adverse unforeseen events without long-term economic, environmental, or social damage. The Australian government will provide support for the development and adoption of techniques including land-capability assessments, farm budgets, marketing plans, and decision support systems.

Research, Education, and Training

The Australian government will make a significant contribution toward programs that (1) provide information, services, and training to assist decision making and (2) implement property management planning. The Australian federal and state governments will develop an integrated national drought research and development program.

DROUGHT MANAGEMENT IN THE 1990s AND BEYOND

The new National Drought Policy proposes that farmers be made as self-reliant as possible in their ability to plan for and manage drought. To this end, it proposed that farmers be made aware of the likely onset of drought through a combination of long-range weather forecasting and estimates of soil moisture levels, plant growth, and senescence using modeling, remote sensing, and Geographic Information Systems (GIS).

Forecasting drought. Long-range climate forecasts encourage farmers to prepare for drought by indicating when the probability of drought occurrence has significantly increased. Forecasts for up to six months ahead have been made possible through our increased understanding of the El Niño phenomenon, in particular the Southern Oscillation Index (SOI) and its relationship to rainfall, especially in northern and eastern Australia (e.g., McBride and Nicholls, 1983; Allan, 1988; Nicholls, 1988 and 1992).

The SOI predictions are most reliable at values more than half a standard deviation from the mean. Given that this situation occurs only about 60% of the time, other predictors are also needed. Furthermore, the SOI does not relate well to rainfall in much of southern and western Australia—hence the current search for additional predictors that could be useful on a regional basis. Sea-surface temperatures from specific ocean areas are showing promise of increasing the ability to predict drought (Nicholls, 1989; Russell, 1991). Variation in synoptic pressure patterns also appears to have some potential in this respect (R. Allan, personal communication).

General circulation (global climatic) models are invaluable tools for evaluating the physical mechanisms contributing to rainfall fluctuations. These models are already demonstrating their ability to simulate droughts, and their associated characteristics, with reasonable accuracy (Gordon and Hunt, 1991; Hunt, 1991). Planned enhancements of these models are expected to further improve their reliability.

Drought monitoring. The Drought Policy Review Task Force (Anonymous, 1990) recommended that priority be given to research into the application of remote sensing and GIS technology to agricultural management systems, including the development of appropriate vegetative and soil moisture indices.

Monitoring vegetation condition by remote sensing. Combined with weather data and current balances of soil water, satellites can monitor the onset of drought, the vegetation's response to drought, and its recovery from the resulting stress.

Satellite data provide details about the condition of the vegetation and the temperature of the earth's surface at levels ranging from the continent to the farm. Contemporary information about the condition of pastures and crops can be derived from the NOAA series of polar-orbiting weather satellites. The reflected red- and near-infrared radiation levels sensed by the NOAA Advanced Very High Resolution Radiometer (AVHRR) can be combined into vegetation indices that can be related to biomass and primary productivity. Images of the entire Australian continent are now available twice a day with a ground resolution (pixel size) of 1.2 km. Lower resolution data transmitted from the NOAA (NOAA APT, 4 km pixels) and GMS (GMS LRFAX, 10 km pixels) meteorological satellites, respectively, may be acquired at relatively low cost (Bryceson and Cannon, 1990).

NOAA AVHRR vegetation indices (such as the Normalized Difference Vegetation Index [NDVI]), which estimate green leaf biomass, provide information on the condition of vegetation. They are already being used for broad-scale monitoring of vegetation condition and greening patterns. In Africa, for example, these data are important in assessing the security of food reserves in areas prone to drought and famine and in contributing to drought prediction. They can provide information on the severity of drought and the termination of drought and response of the vegetation.

In 1986 and 1987, the use of NOAA AVHRR data for drought monitoring and assessment in Queensland was evaluated. The NDVI was found to be well correlated with drought declarations, changes in vegetation condition being indicated that were not readily apparent from sparse Bureau of Meteorology rainfall data (Dudgeon and Jones, 1987).

CSIRO Division of Wildlife and Ecology is developing methods for using AVHRR to monitor spatial greening patterns, length of growing season, and forage biomass in central Australia. This study has been extended to investigate the use of AVHRR data to estimate and forecast both primary (pasture and crop) and secondary (livestock) productivity on a national scale (Foran and Pearce, 1990).

Monitoring soil moisture levels by remote sensing. Impact and severity of drought vary with land use and soil moisture, these being related in part to soil type. A major limitation of using NOAA AVHRR vegetation indices to detect the onset of a drying period is that the vegetation often remains vigorous until the root zone moisture is exhausted; thus an NDVI may remain high when surface moisture has all but disappeared.

Soil moisture can be declining for some weeks before the vegetation is stressed, the sensitivity of the vegetation to the available soil moisture being related to evaporative demand (Denmead and Shaw, 1962). Remote sensing of moisture stress is technically feasible since moisture status is a most important variable determining spectral response from soils and vegetation in both the visible to infrared and microwave spectral regions. The surface temperature of natural materials depends on water/energy interaction. Thermal data provide important information for studies of energy balance and evapotranspiration. Data coverage is available from current sensors at spatial resolutions ranging from a square kilometer to a square meter.

Remotely sensed data from the thermal infrared spectral region are being evaluated for estimating soil moisture budgets and evapotranspiration rates (McVicar et al., 1991). Day and night surface temperatures obtained from NOAA AVHRR data are being related to the amount of water present in the atmosphere, the soil, and the vegetation. The day surface temperature is related to the evaporation of water from the soil, and through vegetation. The day-night surface temperature difference is related to the amount of water being stored near the soil surface. Surface temperature data are particularly sensitive indicators of the onset of a drying sequence, which is the prelude to drought. Data from an airborne multispectral scanner are being used to develop models at a small catchment scale, whereas NOAA AVHRR data are being used to study processes at a large catchment scale.

Estimating Soil Moisture Levels by Water Balance Modeling. Simulation models have been developed within Australia (and overseas) to predict changes in soil moisture over time under pastures and crops. They can distinguish between the response of soils of different water-holding capacities. Such models may be used to advise farmers that soil moisture in the root zone is approaching wilting point before the vegetation itself is clearly stressed.

Regional water balance models can simulate changes in soil moisture over large areas. The result of such modeling is a temporally integrated, spatially sparse plot of rainfall events and the soil moisture. Such results are difficult to validate. Remotely sensed data offers drought researchers an independent means of verifying such modeling. The spatial density of remotely sensed data provides the physical means for interpolating the modeled soil moisture from the spatially sparse network of meteorological stations.

Annual and seasonal herbage accumulation may be estimated and displayed spatially using climatic data and a pasture growth model (Hill et al., 1989). A further development of this approach is embodied in Queensland's SWARD (State-Wide

Analysis of **R**isks of Land **D**egradation) model. The objective of this model is to give farmers prior warning of seasonal conditions that are likely to cause loss of production and land degradation. The model calculates current soil water and pasture yield and predicts an appropriate stocking rate by linking a real-time rainfall network, climatic predictions based on ENSO, an animal numbers data base, and remotely sensed data on a statewide basis. The results will be integrated using Geographic Information Systems (GIS) technology and displayed on maps for use by farmers, agribusiness, and government.

Consideration was recently given to using regular calculations of the Palmer Index (Palmer, 1965; Hatch, 1983) for monitoring drought. This relates regional moisture conditions to the "normal" using a simple water balance model. The United States Department of Agriculture updates the Palmer Index every week during the growing season (April to October) using 11 classes ranging from *above 4.0* (extreme moist spell) to *below -4.0* (extreme drought), with much supporting information. The Index has already been applied to Australian conditions (McDonald, 1984; Smith and Callahan, 1988). However, a recent study of the Palmer Index by Hutchinson et al. (1992) concluded that it offered no advantages over rainfall deciles for estimating the onset and severity of drought.

DEVELOPING DROUGHT MANAGEMENT STRATEGIES FOR FARMS

The new National Drought Policy identifies a need to encourage farm management practices that are consistent with the variable climatic and land capability conditions. In advocating a more self-reliant approach to farm and drought management, the new National Drought Policy recommends that attention be focused on enhancing the ability of farmers to evaluate the most cost-effective management alternatives. Characteristics of self-reliant and efficient drought management therefore need to be defined.

Blackburn et al. (1992) identified strategies for the management of crops, pastures, livestock, and finances that would make farmers more self-reliant. They emphasized budgeting for each season well in advance, identifying stocking and feeding regimes appropriate to their land and business capabilities and risk preferences, and determining the consequences and tactics that would be adopted in the presence and absence of drought. Emphasis was also placed on maintaining a positive mental attitude when faced with drought, since it is more difficult for people to make rational decisions when they and their dependents are being subjected to physical, financial, and emotional stress. Financial strategies identified by Blackburn et al. (1992) as leading to greater self-reliance are summarized in Table 1.

Predrought	Build up cash reserves during good years Budget each year assuming both normal and drought years Stabilize income through off-farm investments
Drought Forecast	Budget for long-range forecast being (a) right, (b) wrong Unload prime or surplus livestock before prices drop Identify least-cost feed supplements Purchase fodder before its value increases Budget to compare feeding and selling strategies Sensitivity analysis of drought duration, feed costs, stock prices Evaluate alternative strategies, e.g., droving, agistment Budget for selling all stock and investing off-farm
Drought	Minimize financial losses to facilitate post-drought recovery Continue comparing feeding and selling strategies Consider raising capital by selling off-farm investments

Table 1. Financial strategies to aid drought preparedness and management.

Land Management

Self-reliant drought management is inextricably linked to the concept of economic and environmental sustainability. Rangelands and improved pastures should be managed so that degradation of soils and vegetation is minimized.

Conditions conducive to soil erosion by wind and water are more prevalent in drought periods (Marshall, 1973). This is caused by the associated reduction in vegetative cover and the drying out of the surface soil. Wind velocity near the ground increases considerably when vegetation is removed. The area of bare ground within a pasture increases with stocking rate, particularly in adverse seasons (White et al., 1980). The decision to retain stock during drought may therefore exacerbate the degradation of vegetative cover (Morley and Daniel, 1992). Further degradation can follow, with soil erosion exacerbated by the action of wind or intensive drought-breaking rains on bare soil.

Self-reliant management therefore requires the choice of an appropriate long-term stocking rate and strategy for grazing management, destocking early in drought (Morley and Daniel, 1992), and possibly planting perennial fodder trees. Only a nucleus of productive and breeding stock should be fed; wethers and steers and the eldest age groups of breeding ewes and cows should be sold or destroyed, depending on the most profitable options. Failed crops can be harvested for grain, cut for hay, or used for grazing. Long-term stocking rates should be both biologically and financially sustainable after allowing for drought. Stocking rates that fail to sustain pastures or viable vegetation are not economical in the long term. In this regard, the sustainability of agricultural systems in some areas may need to be reassessed, given the underlying capacity of the land.

Several studies have been undertaken to assist farmers in identifying optimal management and financial strategies to cope with drought. Least-cost fodder reserves relative to drought incidence and stocking rate have been based on the probability of effective rainfall using an inventory model (Mauldon and Dillon, 1959), and with a simulation model of a grazing system (Thatcher and Lloyd, 1975). An inventory analysis comparing fodder conservation on the farm with purchasing feed grain, as required, was undertaken by Morley and Graham (1971). Options including sale of stock, agistment, and feeding relative to a probability distribution of drought length have been incorporated in a dynamic programming model (Toft and Hanlon, 1979).

MANAGEMENT OF RISK

Risk management is recognized by producers as an important activity in their decision-making process, enabling them to manage their businesses more effectively in a physical environment where drought is a common, though unpredictable, occurrence. Risk management recognizes that producers also operate in an economic environment of less than perfect knowledge. There are three types of risk in agriculture: production, financial, and marketing.

Production risk is imposed primarily by seasonal variability. This risk may be reduced by avoiding excessively high stocking rates, developing strategies for reducing stock numbers in the event of abnormally dry conditions, sowing drought-resistant pasture plants and crops, choosing flock and herd structures and dates of lambing and calving that better relate the nutritional demands of the livestock to the available feed supply, providing shelter for livestock, conserving fodder or growing fodder crops, installing irrigation, and diversifying enterprises.

Strategies that are less risky in terms of production may be much more prone to financial risk. For example, low stocking rates may not allow enough income to be generated in the good seasons to allow a farmer to survive the poor seasons (White, 1987). Stocking according to season may result in stock being purchased at high prices and sold at low prices (Arnold and Bennett, 1975). Dates of lambing or calving that favor production may not favor marketing. Fodder may be conserved on the farm to support high stocking rates, but with the extra stock numbers, less surplus is available to be conserved (Bishop and Birrell, 1975). Irrigation schemes will often be unprofitable even though they reduce the production risk. Diversifying from wool or beef production, for example, into crops or specialist livestock enterprises such as deer or alpaca farming may require substantial capital investment and associated financial risk, and farmers who do diversify often do not have the necessary specialist skills.

Climate predictions may be used to reduce risk. For example, farmers planning to prepare land for sowing crops might not do so, or adjust their inputs accordingly, if they were given an adverse seasonal forecast (Clewett et al., 1991; Hammer and Muchow, 1991). However, a farmer in desperate need of cash to meet financial commitments might sow a crop anyway in the hope that the forecast was incorrect. A farmer might decide to feed a "failing" crop to livestock in the spring based on an adverse forecast. A detailed model of a Merino flock grazing pasture in southeastern Australia (Bowman et al., 1989) showed that farmers who made stocking and selling decisions based on seasonal forecasts could improve their financial performance over the long term, although financial returns were reduced in some years (P. J. Bowman and G. M. McKeon, in preparation). The more accurate the weather forecasts provided in the input to the model, the better the long-term financial performance of the farm.

Decision Support Systems

The conventional approach to risk management is for a farmer to budget for both nondrought and drought years. This may be done using simple spreadsheets describing the expected inputs and outcomes on a farm. These are typically used to budget the cash flow associated with an existing or proposed enterprise(s). The probability of the operation breaking even or earning a required income level for a specific production plan may then be determined. However, seasonal variability and biological feedbacks within a production system are not accounted for in simple budgeting approaches. These require the development and use of decision support systems (DSS).

DSS are aids to decision making. For example, a user may be led through a stepwise sequence of choices until an efficient strategy is decided on. They are usually computer programs with a simulation model as the mathematical core. Properly designed DSS have user-friendly interfaces that more than compensate for their greater internal complexity. DSS models can describe how soil moisture, plant growth, animal intake, growth and production, and the resultant cash flow can vary in response to changes in season and management (e.g., Bowman et al., 1989; McKeon et al., 1990). The models allow the consequences of climatic and financial variations within and between years to be considered. The interactions of plants and livestock with the climate and specific management strategies can also be determined. These interactions often have large economic implications. DSS can therefore make more realistic assessments of risk in relation to a variable climate.

Several research teams have been working around Australia to develop DSS to aid farm and drought management (White and Howden, 1991). These have been shown to be of considerable value in planning to mitigate the harmful effects of drought (Gillard and Monypenny, 1990; Foran and Stafford Smith, 1991; Stafford Smith and Foran, 1992). These systems could be used to determine the likely incidence of adverse seasonal conditions and to identify where existing production practices may be out of equilibrium with such conditions (McKeon and White, 1992).

Reliable DSS can therefore be invaluable aids to farm and drought management. The more variable the climate, the more they are needed. Drought policies and management practices can therefore be tested in different years and locations for different soil and pasture types. This is a technology that is well developed in Australia, with many potential applications to aid drought management and drought policy formulation. Given the diverse nature of the Australian environment and the substantial differences in Australia's grassland ecosystems (namely, annual and perennial improved temperate pastures, semiarid and arid rangelands, tropical and subtropical grasslands), quite different models have been developed to meet the needs of researchers, farmers, and advisers in analyzing such systems.

Strategic farm plans can be readily developed with DSS. Such planning includes decisions such as the setting of appropriate long-term stocking rates of animals in relation to climatic variability and to other characteristics of the local environment that will ensure levels of productivity and profitability consistent with maintaining the soil and vegetation resources. Planning also involves deciding on feeding and stock disposal strategies and making a long-term commitment to financial strategies that reduce the effects of climatic downturns. On extensive properties in the subtropical and arid regions, strategic planning for drought also involves deciding on the number and placement of watering points.

DSS now available in Australia have been enthusiastically received by a small proportion of farmers; most farmers have not yet adopted this technology. For the above "self-reliance" policy to be viable, further development and promotion is needed to enable DSS to meet the drought and farm management needs of individual farmers.

IMPLICATIONS OF CLIMATE CHANGE

Preliminary analyses of the effect of climate change on drought in Australia have been made using output from General Circulation (Global Climate) Models (GCMs) (Gordon et al., 1992; Whetton et al., 1992). These studies indicate that the potential for drought is likely to increase over large areas of the continent as a result of a decline in the number of rain days and an increase in evaporation due to higher temperatures. These studies also indicate that the frequency of days with heavy rainfall will increase. This has serious implications for soil erosion resulting from drought-breaking rainfall. The GCMs are consistently giving strong predictions of changes in rainfall characteristics rather than total changes in rainfall.

The implications of climate change for Australia's agriculture could be considerable. An expected southerly shift in our weather systems would allow tropical crops to be grown further south but could see much of Australia's wheat-sheep zone disappear as the southern areas became even more arid. A southerly migration of tropical pests and diseases of plants and livestock could also occur. International calls for reduced greenhouse gas emissions have important implications for agriculture, given that grazing livestock are the major source of methane emissions in Australia (Howden et al., 1992).

IMPLICATIONS FOR RESEARCH, EXTENSION, AND EDUCATION

Many strategies and tactics are available for farmers to use in order to become more self-reliant. However, a large-scale awareness and education campaign will be needed to ensure that policies based on self-reliance are understood and widely adopted (Blackburn et al., 1992). If farmers are to become self-reliant and better able to manage drought, they also need to understand clearly the circumstances under which they qualify for government assistance. Blackburn et al. (1992) also correctly point out that agricultural research and extension staff, farm management consultants, farm financial counselors, and lecturers at agricultural colleges and universities must be targeted in such education campaigns.

The Bureau of Resource Sciences conducted a national workshop, Drought and Decision Support, on 11-13 March 1992. Developers of decision support systems from around Australia discussed how computer- and satellite-based aids can assist implementation of the new drought policy. Recommendations from this workshop included the need for better prediction, monitoring, and management of drought at the local, regional, and national levels, and the development of systems to allow property and regional managers improved access to forecasting and remote sensing data and decision aids. It was agreed that there should be much more emphasis and education on farm planning and farm business management. It was also agreed that information that could aid drought planning, including climatic, remote sensing, technical, economic, and market data bases and associated software, should reach local farmer and industry groups. This has implications for future private and government infrastructure in terms of both research and extension services and skills and information networks (e.g., White and Shelley, 1992).

Emerging drought prediction and farm management technologies may enable Australian farmers to remain competitive and, at the same time, farm their land in a sustainable way, despite our highly variable climate and fluctuating market forces. The challenge to researchers is the need to develop systems that support reliable, commercially sound decisions and to deliver these systems to farmers in user-friendly packages.

CONCLUSIONS

The adoption of a new national drought policy to protect and manage Australia's natural resources effectively, given a highly variable climate, is a major step forward in reducing the impact of drought. Management strategies that lead to degradation of soils and vegetation will no longer be rewarded through inappropriate systems for providing drought relief. Industry and government will emphasize the identification and adoption of farming systems and management practices that are biologically, physically, and financially sustainable, and monitor these accordingly.

The adoption of whole-farm planning will provide the key means by which risk management, productivity growth, and sustainable development will be achieved. This will be aided by computer- and satellite-based technologies that have the potential to aid climate prediction, land monitoring, and physical and financial management of farms. Farmers will also need substantial education on all aspects of effective, self-reliant drought management.

ACKNOWLEDGMENTS

The authors thank Drs. K. P. Bryceson and R. Johnston of the Bureau of Resource Sciences; Dr. F. H. W. Morley of the School of Veterinary Science, University of Melbourne; and Dr. D. I. Smith of the Centre for Resource and Environmental Studies, Australian National University, for their useful comments on the manuscript.

ABOUT THE AUTHORS

David White is a senior principal research scientist in the Bureau of Resource Sciences, within the Australian Department of Primary Industries and Energy (DPIE). His contributions on risk management and decision support have been incorporated into national policies on drought management, agricultural meteorology, and sustainable agriculture. Before joining the Bureau, he worked for more than 21 years in the Victorian Department of Agriculture, initially servicing the sheep industry in northern Victoria and later simulating and analyzing the management of agricultural systems at the Animal Research Institute at Werribee. Dr. White is principal editor of *Agricultural Systems and Information Technology* and a fellow of the Australian Institute of Agricultural Science.

Drew Collins is manager of the Land and Forestry Economics Section within the Australian Bureau of Agricultural and Resource Economics (ABARE, within DPIE). He is a research economist with a research interest in public policy affecting Australia's forestry, fishery, and agricultural industries. His research has focused on policies to promote the adoption of sustainable agricultural practices and the development of efficient resource industries. Recently this has included research into water and forestry pricing and allocation systems and the economics of alternative farm practices.

Mark Howden completed his Ph.D. on the ecology of tropical grassland systems. He then worked on drought and rangeland management within the Queensland Department of Primary Industries. He is now a senior research scientist in the Bureau of Resource Sciences and is working with teams researching issues relating to climate variability and climate change. This research has shown that both droughts and unusually wet years can be predicted with reasonable accuracy over much of eastern Australia, and it has also demonstrated how these predictions can be integrated into grazing and cropping management systems to reduce risk, increase profits, and increase the sustainability of these enterprises.

REFERENCES

Allan, R. J. 1988. El Niño-Southern Oscillation influences in the Australasian region. *Progress in Physical Geography* 12:313-49.

Anderson, J. R.; and J. B. Hardaker. 1973. Management decisions and drought. *In* J.V. Lovett, ed. *The Environmental, Economic and Social Significance of Drought*; pp. 220-44. Angus and Robertson, Sydney.

Anonymous. 1990. *National Drought Policy Final Report*. Drought Policy Review Task Force. Australian Government Publishing Service, Canberra.

Arnold, G. W.; and D. Bennett. 1975. The problem of finding an optimum solution. *In* G.E. Dalton, ed. *Study of Agricultural Systems*; pp. 129-73. Applied Science Publishers, London.

Bishop, A. H.; and H. A. Birrell. 1975. Effect of stocking rate, fodder conservation and grazing management on the performance of wether sheep and pastures in south-west Victoria. 2. Seasonal wool growth rate, liveweight and herbage availability. *Australian Journal of Experimental Agriculture and Animal Husbandry* 18:41-50.

Blackburn, A.; S. Inglis; L. Polkinghorne; R. Ashby; and C. Hacking. 1992. *Characteristics of Self-Reliant and Efficient Drought Management*. Report to Bureau of Resource Sciences.

Bowman, P. J.; D. A. Wysel; D. G. Fowler; and D. H. White. 1989. Evaluation of a new technology when applied to sheep production systems: Part I—Model description. *Agricultural Systems* 29:35-47.

Bryceson, K. P.; and R. Cannon. 1990. Acquisition and use of low-cost meteorological satellite data in the Australian Plague Locust Commission. *Computers and Electronics in Agriculture* 5:47-64.

Campbell, R.; P. Crowley; and P. Demura. 1983. Impact of drought on national income and employment. *Quarterly Review of the Rural Economy* 5:254-57. Bureau of Agricultural Economics, Canberra.

Clewett, J. F.; S. M. Howden; G. M. McKeon; and C. W. Rose. 1991. Optimising farm dam irrigation in response to climatic risk. *In* R. C. Muchow and J. A. Bellamy, eds. *Climatic Risk in Crop Production: Models and Management for the Semiarid Tropics and Subtropics*; pp. 307-28. Wallingford (CAB International).

Crean, S. 1992. *New national drought policy*. Media release, Minister for Primary Industries and Energy, 20 August 1992, DPIE92/66C, Parliament House, Canberra, Australia.

Denmead, O. T.; and R. H. Shaw. 1962. Availability of soil water to plants as affected by soil moisture content and meteorological conditions. *Agronomy Journal* 54:385-90.

Dudgeon, G. S.; and S. Jones. 1987. Queensland drought assessment using AVHRR. *Proceedings of the 4th Australasian Remote Sensing Conference*, Adelaide, August; pp. 160-69.

Foran, B.; and G. Pearce. 1990. The use of NOAA AVHRR and the normalized difference vegetation index to assess the 1988/89 summer growing season in central Australia. *Proceedings of the 5th Australasian Remote Sensing Conference*, Perth, October; pp. 298-307.

Foran, B.; and D. M. Stafford Smith. 1991. Risk, biology and drought management strategies for cattle stations in Central Australia. *Journal of Environmental Management* 33:17-33.

Gentilli, J. 1971. The main climatological events. *In* J. Gentilli, ed. *World Survey of Climatology*. Volume 13, *Climates of Australia and New Zealand*; pp. 119-88.

Gibbs, W. J.; and J. V. Maher. 1967. Rainfall deciles as drought indicators. *Bureau of Meteorology Bulletin* 48:118, Bureau of Meteorology, Melbourne.

Gillard, P.; and R. Monypenny. 1990. A decision support model to evaluate the effects of drought and stocking rate on beef cattle properties in northern Australia. *Agricultural Systems* 34:37-52.

Gordon, H. B.; and B. G. Hunt. 1991. Droughts, floods, and sea-surface temperature anomalies: A modelling approach. *International Journal of Climatology* 11:347-65.

Gordon, H. B.; P. H. Whetton; A. B. Pittock; A. M. Fowler; and M. R. Haylock. 1992. Simulated changes in daily rainfall intensity due to the enhanced greenhouse effect: Implications for extreme rainfall events. *Climate Dynamics* (in press).

Hammer, G. L.; and R. C. Muchow. 1991. Quantifying climatic risk to sorghum in Australia's semiarid tropics and subtropics: Model development and simulation. *In* R. C. Muchow and J. A. Bellamy, eds. *Climatic Risk in Crop Production: Models and Management for the Semiarid Tropics and Subtropics*; pp. 205-32. Wallingford (CAB International).

Hastings, P. A. 1990. Southern Oscillation influences on tropical cyclone activity in the Australia/south west Pacific region. *International Journal of Climatology* 10:291-98.

Hatch, W. L. 1983. Selective guide to climatic data sources. *Key to Meteorological Records Documentation No. 4.11*. NOAA/NESDIS, U.S. Department of Commerce, Washington, D.C.

Heathcote, R. L. 1991. Managing the droughts? Perception of resource management in the face of the drought hazard in Australia. *Vegetatio* 91:219-30.

Hill, M. J.; M. F. Hutchinson; and A. C. Kirby. 1989. Mapping herbage accumulation for the Hunter region of Australia using climatic analysis, growth indices and field experiments. *Proceedings of the XVIth International Grassland Congress*, Nice, France; pp. 1369-70.

Howden, S. M.; G. M. McKeon; and I. E. Galbally. 1992. Methane emissions from Australian agriculture: Other sources. *Australian Institute of Agricultural Science (Queensland)* (in press).

Hunt, B. G. 1991. The simulation and prediction of drought. *Vegetatio* 91:89-103.

Hutchinson, M. F.; R. MacArthur; and D. I. Smith. 1992. Short summary of analysis of climatic drought. *In* Bureau of Resource Sciences (Workshop Report). *Drought and Decision Support* (in press).

Marshall, J. K. 1973. Drought, land use and soil erosion. *In* J. V. Lovett, ed. *The Environmental, Economic and Social Significance of Drought*; pp. 55-77. Angus and Robertson, Sydney.

Mauldon, R. G.; and J. L. Dillon. 1959. Droughts, fodder reserves and stocking rates. *Australian Journal of Agricultural Economics* 3:45-47.

Morley, F. H. W.; and G. Daniel. 1992. Drought, stocking rate and soil loss. *Proceedings of the Australian Society of Animal Production* 19:323-25.

Morley, F. H. W.; and Graham, G. Y. 1971. Fodder conservation for drought. *In* J. B. Dent and J. R. Anderson, eds. *Systems Analysis in Agricultural Management*; pp. 125-41. Elsevier, Amsterdam.

McBride, J. L.; and N. Nicholls. 1983. Seasonal relationships between Australian rainfall and the Southern Oscillation. *Monthly Weather Review* 111:1998-2004.

McDonald, N. S. 1984. *Palmer Drought Index, N.S.W. Meteorological Districts 1913-1982*. Department of Geography Occasional Paper No. 10, The Australian National University.

McKeon, G. M.; and D. H. White. 1992. El Niño and better land management. *Search* 23:197-200.

McKeon, G. M.; K. A. Day; S. M. Howden; J. J. Mott; D. M. Orr; W. J. Scattini; and E. J. Weston. 1990. Management for pastoral production in northern Australian savannas. *Journal of Biogeography* 17:355-72.

McVicar, T. R.; D. L. B. Jupp; J. Walker; and G. Tian. 1991. Remote sensing with physical models of soil moisture status to monitor land degradation and drought. *Proceedings of the 12th Asian Conference on Remote Sensing*, Singapore, Oct. 30-Nov. 5; pp. A-1-2-1–A-1-2-6.

Nicholls, N. 1988. El Niño-Southern Oscillation and rainfall variability. *Journal of Climate* 1:418-21.

Nicholls, N. 1989. Sea surface temperatures and Australian rainfall. *Journal of Climate* 2:965-73.

Nicholls, N. 1992. ENSO, drought and flooding rain in southeast Asia. *In* H. Brookfield. *Toward a Sustainable Environmental Future for the Southeast Asian Region.* United Nations University, Tokyo (in press).

Palmer, W. C. 1965. *Meteorological Drought.* Research Paper No. 45, U.S. Weather Bureau, Washington, D.C.

Peters, A. J.; D. C. Rundquist; and D. A. Wilhite. 1991. Satellite detection of the geographic core of the 1988 Nebraska drought. *Agriculture and Forest Meteorology* 57:35-47.

Purtill, A.; M. Backhouse; A. Abey; and S. Davenport. 1983. A study of the drought. Supplement to *Quarterly Review of the Rural Economy* 5(1). Bureau of Agricultural Economics, Canberra.

Reynolds, R.; D. Watson; and D. Collins. 1983. *Water Resource Aspects of Drought.* Australian Government Printing Service, Canberra.

Russell, J. S. 1991. Prospects for incorporation of long-term weather forecasting into crop decision support systems. *In* R. C. Muchow and J. A. Bellamy, eds. *Climatic Risk in Crop Production: Models and Management for the Semiarid Tropics and Subtropics*; pp. 467-87. Wallingford (CAB International).

Smith, D. I.; and S. D. Callahan. 1988. *Climatic and Agricultural Drought, Payments and Policy: A Study of New South Wales.* CRES Working Paper 1988/16, Centre for Resource and Environmental Studies, Australian National University.

Stafford Smith, M.; and B. Foran. 1992. An approach to assessing the economic risk of different drought management tactics on a South Australian pastoral sheep station. *Agricultural Systems* 39:83-105.

Thatcher, L. P.; and A. G. Lloyd. 1975. Least-cost feed reserve using drought probabilities derived from a grazing model. *Review of Marketing and Agricultural Economics* 43:113-25.

Toft, H. I.; and P. W. O'Hanlon. 1979. A dynamic programming model for on-farm decision making in a drought. *Review of Marketing and Agricultural Economics* 47:5-16.

Troup, A. J. 1965. The Southern Oscillation. *Quarterly Journal of the Royal Meteorology Society* 91:490-506.

Whetton, P. H.; A. M. Fowler; M. R. Haylock; and A. B. Pittock. 1992. Implications of climate change due to the enhanced greenhouse effect on floods and droughts in Australia. *Climatic Change* (submitted).

White, D. H. 1987. Stocking rate. *In* R. W. Snaydon, ed. *Managed Grasslands. B. Analytical Studies*; pp. 227-38. Elsevier, Amsterdam.

White, D. H.; and S. M. Howden. 1991. The emerging role of decision support systems in managing for drought on Australian farms. *In* J. Stuth and B. Lyons, eds. *Decision Support Systems for Resource Management*; pp. 129-32. Texas A&M University, College Station, Texas, U.S.A.

White, D. H.; and S. M. Howden. 1992. Drought: Reducing the impact. *Interface* 1(2):4-9; Bureau of Resource Sciences, Canberra.

White, D. H.; and E. P. Shelley. 1992. The role of information and information networks in decision support: The Australian experience. *In* J. Stuth and B. Lyons, eds. *Emerging issues for Decision Support Systems for Management of Grazing Lands*. UNESCO-MAB Book Series (in press).

White, D. H.; B. J. McConchie; B. C. Curnow; and A. H. Ternouth. 1980. A comparison of levels of production and profit from Merino ewes and wethers grazed at various stocking rates in northern Victoria. *Australian Journal of Experimental Agriculture and Animal Husbandry* 20:296-307.

Chapter 13
Drought Mitigation in the United States: Progress by State Government

Donald A. Wilhite and Steven L. Rhodes

INTRODUCTION

Drought has been a recurrent feature of the American landscape in recent years, resulting in significant impacts in many economic sectors, including agriculture, transportation, energy, recreation, and health; it has also had adverse environmental consequences. For example, the economic impacts of the 1976-77 and 1988 droughts have been estimated at nearly $35 billion and $40 billion, respectively (NOAA, 1982; Riebsame et al., 1990). Other drought years, such as 1980, 1983, 1986, and 1989-91, resulted in significant losses as well. The present and future impacts of the current (1992) drought in the western states and in portions of the east are likely to be substantial and long-lasting. Almost without exception, the occurrence of widespread severe drought in the United States has illustrated the low level of drought preparedness that has existed in federal and state governments. Assessment and response programs that were implemented during the 1970s have been characterized as largely ineffective, poorly coordinated within and between levels of government, and untimely (GAO, 1979; Wilhite et al., 1986). Although state government has made considerable progress in drought preparedness, the verdict is still out on whether the federal government's response to the 1988-89 drought had improved significantly over previous response efforts. Riebsame et al. (1990) suggests only a moderate improvement in efficiency. The lessons of past response efforts in the United States strongly suggest that a "risk management" or proactive approach to drought management would be a more effective mitigation tool than the "crisis management" or reactive approach heretofore practiced.

The primary purpose of this chapter is to discuss the growth in state drought contingency plans (DCPs) in the United States and to document their effectiveness as a mitigation and response tool during recent periods of water shortage. The chapter also presents a rationale for drought planning and the principal components or elements of existing state plans. Recognizing that drought is the primary but not the only stimulus for drought plan development in most political settings, we also speculate on other factors that have fostered the increased interest of state governments in drought planning. The chapter concludes by proposing some directions for state and federal drought planning efforts in the United States. The underlying assumption is that the growth of state drought contingency planning efforts in the United States has implications for the international scientific and policy community. As interest in drought preparedness increases worldwide, the lessons learned in the United States and other countries with an extensive history in drought response may assist policy makers in understanding the social, economic, environmental, and political forces that influence the planning process.

RATIONALE FOR DROUGHT PREPAREDNESS

The reasons behind the development of a DCP by state government in the United States have been addressed elsewhere (Wilhite, 1991a; see also Chapter 6). Simply stated, sharply focused contingency plans, prepared in advance, can assist government and others in the early identification of drought and its likely impacts; improve the efficiency of resource (fiscal, human, and natural) allocation; lessen personal hardship; and, ultimately, reduce drought-related impacts, the need for government-sponsored assistance programs, and long-term vulnerability to climate-induced periods of water shortage. In the absence of a plan, communication within and between agencies and levels of government is often poorly developed, which in turn leads to untimely or inappropriate decisions (Wilhite et al., 1986). Drought contingency plans improve the coordination and efficiency of assessment and response actions of the wide range of state agencies with responsibilities for water and also improve the linkages and working relationships with the myriad of federal agencies that are represented in each state. In the long term, contingency plans can reduce societal vulnerability to periods of water shortage and the need for drought assistance from the federal government.

Drought planning has been defined as actions taken by individual citizens, industry, government, and others in advance of drought for the purpose of mitigating some of the impacts and conflicts associated with its occurrence (Wilhite et al., 1986; Wilhite, 1991a). To be successful, drought planning must be integrated between the national and state levels of government, involving existing regional (multistate) organizations as well as the private sector where applicable. Examples of regional organizations in the United States that have been active in drought planning include the Great Lakes Commission, Western States Water Council, and Delaware River Basin Commission.

STATE DROUGHT CONTINGENCY PLANS: TRENDS AND CHARACTERISTICS

Progress by states in the development of DCPs in recent years has been extraordinary. During the widespread and severe drought of 1976-77, for example, no state had prepared a formal drought strategy (Wilhite, 1991a). In 1982 only three states had developed plans: South Dakota (1981), Colorado (1981), and New York (1982). Typically, states have relied on the federal government to come to their rescue with massive relief programs when water shortages reach near-disaster proportions. The federal government provided nearly $8 billion in relief in response to the sequence of drought years in the mid-1970s; federal assistance efforts totaled more than $5 billion in response to the 1988-89 droughts (Wilhite et al., 1986; Riebsame et al., 1990). This assistance was provided to drought victims through an assortment of federal programs administered by numerous federal agencies. In 1977, forty separate programs were administered through sixteen agencies (GAO, 1979).

During the past decade, an additional twenty-four states have developed and implemented formal drought contingency plans (see Figure 1). Twenty-seven states now have plans. In addition, three states are developing or have expressed an intent to develop plans in the near term. These planning efforts have often been conducted in conjunction with a state's overall water management planning initiative. Clearly, states can now be labeled policy innovators in planning for drought (Wilhite, 1990).

Because of the many resources that are now available to assist governments in the drought planning process, it is likely that the number of states with plans will continue to increase (but at a slower pace) and that existing plans will be revised periodically to

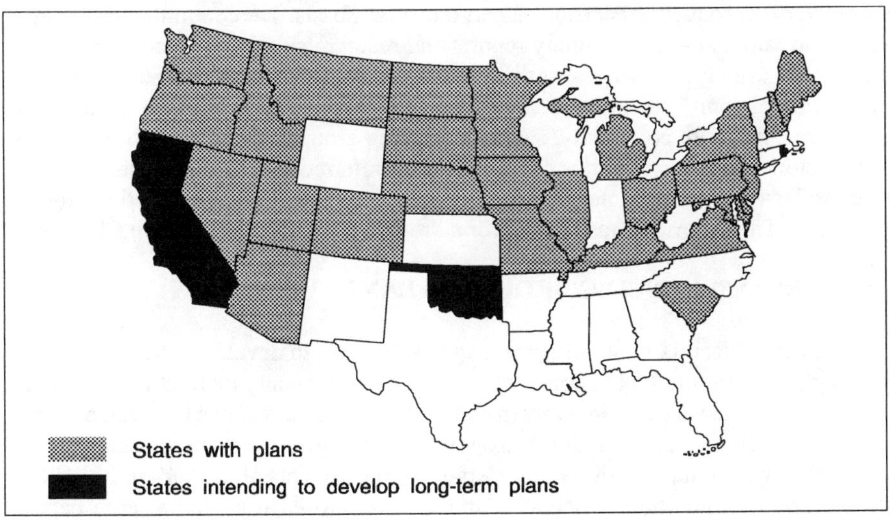

Figure 1. Status of drought planning in the United States, 1992 (Wilhite, 1992).

incorporate new mitigation technologies as they become available. Model plans (Western States Water Council, 1987; Wilhite, 1990 and 1991a) and 27 state plans now provide a critical reference for states desiring to develop a plan or revise an existing plan. In addition, several regional organizations have considerable experience in drought planning and can assist states in plan development (e.g., Delaware River Basin Commission, Great Lakes Commission, Western States Water Council). At the same time, some federal agencies have attempted to improve their drought management capability but seem reluctant to pursue the development of a coordinated national policy and plan.

An examination of existing state drought plans reveals that they have certain key elements in common (Wilhite, 1991b). Administratively, a task force is responsible for the operation of the system and is directly accountable to the governor. The task force keeps the governor advised of water availability and potential problem areas; it also recommends policy options for consideration. Operationally, drought plans have three features in common. First, a water availability committee is established to continuously monitor water conditions and prepare outlooks a month or season in advance. Since most of the information necessary to comprehensively monitor water conditions (i.e., precipitation and temperature, streamflow, groundwater levels, snowpack, soil moisture, meteorological forecasts) is available from state or federal agencies, the primary role of the committee is to coordinate the collection and analysis of this information and the delivery of products to decision makers on a timely basis. The committee assimilates this information and issues timely reports and recommendations. Second, a formal mechanism usually exists to assess the potential impacts of water shortages on the most important economic sectors. In some states this task is accomplished by a single committee, or, more commonly, separate working groups are established to address each sector. Third, a committee or the task force referred to previously usually exists to consider current and potential impacts and recommend response options to the governor. The generic infrastructure for a drought plan was discussed in Chapter 6.

FACTORS INFLUENCING DROUGHT PLAN DEVELOPMENT

Numerous factors may influence a state's decision to develop a drought contingency plan. It is important to note at the outset that the primary motivating factors for plan development vary considerably from state to state. The decision to develop (or not to develop) a plan is a highly individual one that, in reality, reflects several factors, some of which may be unique to that state. Certainly, the droughts of the mid- to late 1980s and early 1990s and their associated impacts seem to have played a large part in generating momentum at the state level. Coupled with this factor has been a growing awareness of and concern about the inefficiency of past response efforts and a genuine desire on the part of state government to improve their assessment and response capability. As Figure 1 demonstrates, however, the pattern of drought plans currently in existence cannot be explained on the basis of climatological factors alone since many states with plans lie in the more humid eastern portion of the country, a region that historically has had a lower probability of severe drought occurrence. When droughts occur in this region, they tend to be of shorter duration (Karl, 1983).

A typology of state behavior was constructed to help explain the current pattern of drought contingency plan development and its relationship to the occurrence of drought during the past ten years. Lester (1986) developed this methodological approach to decipher the relationships between state commitment to environmental quality and state dependency on federal aid. Lester's typology was modified for this study to attempt to explain state commitment to drought plan preparation on the basis of the frequency of severe to extreme drought during the period from 1982 to 1991. The classification of severe to extreme drought was based on the Palmer Drought Severity Index (PDSI), a meteorological drought index used extensively by federal and state government in the United States to monitor long-term climate trends, including the pattern and severity of drought (Palmer, 1965). The PDSI has been used routinely by the federal government as one of the principal criteria for disaster designation (i.e., eligibility to receive federal drought relief; Wilhite et al., 1986). Alaska and Hawaii were not included in this study because PDSI values were unavailable for comparison.

The 1982-91 period was chosen to explain the relationship between state commitment to drought plan preparation and drought frequency because of the rapid growth in the number of plans developed during these years. To arrive at an average statewide drought frequency value for the period in question, the frequency of severe to extreme drought was first determined for each climatological division (i.e., relatively homogeneous climatic regions). Most states contain between six and nine climatic divisions. The divisional averages were then averaged to derive an unweighted state average. States were then divided into two drought frequency groups, ≥10 % (High) and <10 % (Low). The 10% threshold was chosen because a one-in-ten probability of severe to extreme drought is often considered significant by decision makers (e.g., High Plains Associates, 1982).

State commitment to drought plan preparation was divided on the basis of whether or not a plan had been developed (Yes or No). Thus, states with a high frequency of severe to extreme drought during the 1982-91 period and a drought plan would be placed in Sector A. Conversely, states with a relatively low frequency of severe to extreme drought and no drought plan would be placed in Sector D. The outcome of this typological analysis is presented in Figure 2.

A significant number of states (15) were classified in Sector A (high drought frequency and a drought plan or intent to develop one). Most of these states are in the west, where drought is a common feature of the climate. Only South Dakota had developed a plan before 1983. This plan was developed primarily in response to the mid-

State Commitment to Drought Plan Preparation

Frequency of severe to extreme drought (PDSI ≤ -3.0) during 1983-89	High (Drought Plan)	Low (No Drought Plan)
High (≥ 10%)	**A** California Idaho Montana Nebraska North Dakota Oregon Florida Nevada Utah Washington Delaware South Dakota Iowa Arizona Minnesota	**C** Wyoming
Low (< 10%)	**B** Kentucky Maine New Hampshire Ohio Colorado New Jersey Missouri Oklahoma Pennsylvania South Carolina New York Virginia Illinois Maryland Michigan Rhode Island	**D** Connecticut Georgia Indiana Louisiana Massachusetts Mississippi Kansas Vermont North Carolina Wisconsin Tennessee West Virginia Alabama Arkansas New Mexico Texas

Figure 2. A typology of state behavior: Frequency of severe to extreme drought during 1983-89 and commitment to drought plan preparation.

1970s droughts. In the more humid east, only Delaware falls into this sector, experiencing severe to extreme drought in 10% of the months during the 1982-91 period. The only other eastern state included in Sector A is Florida, with a 12.4% frequency. Florida is unique in that the authority for water management/drought planning is transferred to water management districts.

Fifteen states fall into Sector D (low drought frequency and no drought plan). The majority of these states are located in the more humid east. Three states in this sector that might be expected to have a plan (based on their climatic regimes) but do not are Kansas, New Mexico, and Texas. These states are all located in the drought-prone west. However, in each of these cases, the statewide average percent frequency of severe to extreme drought was below the 10% threshold during the 1982-91 period. Drought frequency greater than 10% was associated with portions of Kansas and Texas but averaged out to less than 10% for the state as a whole.

The states classified in sectors B and C are much less easily explained on the basis of climatology. In Sector B (states with a low drought frequency and a drought plan or intent to develop one), sixteen states are represented, all but two of which are located in the east. Colorado developed a plan in 1981 in response to two significant droughts that occurred in 1977 and 1980. Governor Richard Lamm was in office during both of these events and witnessed the inefficiency of assessment and response efforts (Wilhite, 1990). In 1981, he requested the Department of Disaster Emergency Services to develop a drought plan to enhance state preparedness and future response. Thus, the existence of the Colorado plan is directly attributable to drought, but in the period before 1982. Oklahoma has expressed an intent to develop a drought plan, although the frequency of severe drought occurrence during 1982-91 is very low (1%). This frequency is unusually low; severe drought occurred 12.3% of the time during 1951-91. New York developed a plan in 1982 in response to numerous factors, including the potential effects of drought on a water supply system stretched by the demands of an enormous urban population. Other factors that may help to explain the development of plans in the remaining thirteen states will be discussed in the next section of this paper.

Only one state appears in Sector C (high drought frequency and no drought plan)—Wyoming. Statewide drought frequency during the 1982-91 period was quite high in Wyoming (33.9%). Climatology alone would certainly suggest that this state would have developed a plan to address assessment and response issues directly related to periods of water shortage. The absence of a drought contingency plan in Wyoming may be attributed to one or more factors, including its small population base, low population density, reliance on the prior appropriation doctrine for allocating water during water-short periods, and political environment.

Although climate has been a primary stimulus to drought plan development in the past ten years, other factors, such as increasing population pressure, conflicts between water users, and antiquated water supply systems, at times in conjunction with climate, have motivated states to develop plans. Another primary motivating factor has been the policy environment of the 1980s. Some of the major policy-related factors that deserve consideration are reviewed briefly below.

DROUGHT PLANNING AND THE 1980s POLICY ENVIRONMENT

How did the policy environment of the 1980s contribute to the large number of states (thirty in Sectors A and B, including those with low frequency of severe to extreme drought during 1982-91) developing drought contingency plans over the past decade? It is difficult to identify specific decisions, experiences, or other factors that prompted so many states to undertake the preparation of a formal drought plan. State governments for the most part do not maintain permanent drought response offices with stable staffing, and state employees routinely move to new responsibilities within government or to the private sector. Thus, institutional memory is often quite short (i.e., lessons from recent droughts are soon forgotten).

A general assessment of the public policy environment during the 1980s suggests several policy themes that may have contributed to a growing awareness and understanding of the complexities of drought management. These themes may have directly or indirectly affected the behavior of state government officials in pursuing the development of a drought plan. These themes raise philosophical and practical questions about the nature of federal-state relations and particular water resource planning and management innovations that were adopted by several states during the 1980s.

The issue of federal-state relations and their possible influence on state drought planning involves three distinct elements: (1) improved capabilities of state governments in conjunction with the Reagan administration's "new federalism" initiative and concurrent federal regulatory mandates to state and local governments; (2) states' concerns about federal intrusion into state-level water resource planning and water rights; and (3) some states' early experiences in working with the new Federal Emergency Management Agency in the early 1980s. In the case of several states, a fourth incentive to plan for droughts may have developed as a consequence of policy innovations regarding water supply planning and management. Two specific examples are: (1) the large number of states adopting groundwater management policies during the 1980s, and (2) recent water supply management problems that have been exacerbated by the riparian doctrine of water law that predominates in the states east of the Mississippi River. Although it cannot be proved that all or any of these influences affected drought planning in any particular state, it is intriguing that these perturbations to federal-state relations and growing government concerns about future water supplies occurred during the period of accelerated state adoption of drought plans.

It is hypothesized that issues such as these may have contributed to an increased awareness of the value of drought planning within some state governments. Coupled with actual drought experiences or knowledge of other states' drought impacts, these policy influences may have helped create conditions conducive to increased state attention to drought planning.

STATE MITIGATION ACTIONS DURING RECENT DROUGHTS: SELECTED EXAMPLES

As previously stated, drought plans are developed to reduce drought-related impacts and improve efficiency in the allocation and use of resources. States were surveyed recently to determine mitigation actions implemented in response to the widespread occurrence of severe drought that has affected large portions of the United States during the period from 1986 to 1992 (Wilhite, 1992). At this writing, severe drought conditions continue throughout most of the western states and in selected areas in the central and eastern portion of the country.

Hy and Waugh (1990) have defined mitigation as "activities that reduce the degree of long-term risk to human life and property from natural and man-made hazards." Mitigative actions were purposely not defined in the survey instrument referred to above—states were given the flexibility to define mitigation by including those actions or activities that they felt were appropriate. Those mitigation activities identified by states and/or local municipalities during recent droughts were diverse, reflecting regional differences in impacts, legal and institutional constraints, and characteristics of contingency plans. The diversity in responses was also related to the wide range of state agencies with principal authority for planning and mitigative actions (e.g., agriculture, natural resources, water resources, emergency or disaster management).

Mitigation actions can be clustered into nine primary areas, as shown in Table 1. These actions represent the full range of possible mitigative actions, from monitoring and assessment programs to the development of drought contingency plans. Some of the actions included were adopted by many states, but others may have been adopted only in a single case. One of the most innovative and successful mitigation actions implemented during recent droughts was the water bank program initiated by California (California Department of Water Resources, 1992; Howitt et al., 1992). This program is discussed in more detail in Chapter 14. Actions listed in Table 1 include, for the most part, those that were adopted by state government. Mitigative actions adopted by other levels of government and the private sector are not summarized in Table 1. However, they would, in most cases, be included in these nine categories.

Many of the mitigative programs implemented by states during recent droughts can be characterized as emergency or short-term actions taken to alleviate the crisis at hand. However, these actions were quite successful. Other activities, such as legislative

Category	Specific Action
Assessment Programs	Developed criteria or triggers for drought-related actions Developed early warning system, monitoring program Conducted inventories of data availability Established new data collection networks Monitored vulnerable public water suppliers
Legislation/ Public Policy	Prepared position papers on public policy issues Examined water rights statues for possible modification during water shortages Passed legislation to protect instream flows
Water Supply Augmentation/ Development of New Supplies	Issued emergency permits for water use Provided pumps and pipes for distribution Proposed and implemented program to rehabilitate reservoirs to operate at design capacity Undertook water supply vulnerability assessments Inventoried self-supplied industrial water users for possible use of their supplies for emergency public water supplies Inventoried and reviewed reservoir operation plans
Public Awareness/ Education Programs	Organized drought information meetings for the public and the media Implemented water conservation awareness programs Published and distributed pamphlets to individuals, businesses, and municipalities on water conservation techniques and agricultural drought management strategies Organized workshops on special drought-related topics Prepared sample ordinances on water conservation for municipalities and domestic rural supplies
Technical Assistance on Water Conservation	Provided advice on potential new sources of water Evaluated water quantity and quality from new sources Advised water suppliers on assessing vulnerability of existing supply system Recommended the adoption of water conservation measures to suppliers

Table 1. Drought-related mitigative actions.

Category	Specific Action
Demand Reduction/ Water Conservation Programs	Established stronger economic incentives for private investment in water conservation
	Encouraged voluntary water conservation
	Improved water use and conveyance efficiencies
	Implemented water metering and leak detection programs
Emergency Response Programs	Established alert procedures for water quality problems
	Stockpiled supplies of pumps, pipes, water filters, and other equipment
	Established water hauling programs for livestock from reservoirs and other sources
	Compiled list of locations for livestock watering
	Established hay hotline
	Provided funds for improvement of water systems, developing new systems, and digging of wells
	Provided funds for recovery programs for drought and other natural disasters
	Lowered well intakes on reservoirs for rural water supplies
	Extended boat ramps and docks in recreational areas
Water Use Conflict Resolution	Acted to resolve emerging water use conflicts
	Negotiated with irrigators to gain voluntary restrictions on irrigation in areas where domestic wells were likely to be affected
	Established a water banking program
	Clarified state law regarding sale of water
	Clarified state law on changes in water rights
	Suspended water use permits in watersheds with low water levels
	Investigated complaints of irrigation wells interfering with domestic wells
Drought Contingency Plans	Recommended to water suppliers the development of drought plans
	Established state-wide contingency plan
	Evaluated worst-case drought scenarios for possible further actions

Table 1 (contd.).

actions, drought plan development, and the development of water conservation and other public awareness programs, are considered as actions with a longer-term impact. As states gain more experience assessing and responding to drought, future actions will undoubtedly become more timely and effective and less reactive. Although the mitigative actions taken by states in response to recent drought conditions are numerous when considered in total, individual state actions were in most cases quite narrow. In the future, state drought contingency plans must become broader in scope, addressing a broader range of potential mitigative actions, including provisions for expanding the level of intergovernmental coordination. To be successful, drought planning must be integrated between local, state, and federal levels of government and with regional organizations, as appropriate. In time this will help states avoid or reduce the impacts, conflicts, and personal hardship.

CONCLUSIONS AND RECOMMENDATIONS

This chapter has reviewed climatological and nonclimatological factors that may have contributed to the adoption of state drought response plans in the United States since 1982. Although drought experiences and concerns about possible impacts of drought probably have had a more direct effect on state drought response planning, the policy environment of the past decade illustrates additional factors that may have contributed to increased awareness within state governments of the potential value of being prepared to respond to drought.

Of course, for those states that have not adopted formal drought response plans, specific influences have also affected their decision-making processes. Yet the sheer number of states moving forward with formal drought plans in the past decade prompts questions about the reasons for this wave of apparent drought awareness. Our assessment suggests a picture of drought policy formulation that is much more complex and multidimensional than can be accounted for by climatological explanations alone.

Given progress made at the state level in the past decade as a consequence of drought experiences and perhaps the other influences discussed in this chapter, what drought planning initiatives should be of highest priority for state (and federal) government in the future? First, we believe that there must be greater integration of existing drought management and planning activities with existing water management and natural hazard management and planning activities at both levels of government. This problem stems largely from the unique character of drought, a slow-onset disaster, in contrast to quick-onset natural disasters such as earthquakes, hurricanes, and floods. Typically, drought planning has focused on the mitigation of impacts, a shorter-term objective, rather than on longer-term water management and planning issues. At the federal level, the activities of the numerous agencies with drought-related missions in assessment, management, mitigation, and planning have never been integrated to provide a more coherent approach to drought management. A national policy and plan emphasizing a more proactive risk management approach, as called for by many agencies and organizations, must be developed.

Second, it is recommended that the mitigation, response, and planning efforts of state and federal government be better coordinated. Drought is a distinctly regional phenomenon with national and international implications. Therefore, linkages should exist between those agencies in each state with primary authority for ongoing drought planning activities. Networking between states will facilitate the dissemination of new mitigation technologies. This network will result in greater interaction between state and federal agencies on drought-related issues.

Regional organizations can play a vital role in linking the assessment and response activities of states within their region and can also help to establish dialogue with federal agencies. It is important to note here the need to coordinate drought-related planning activities at the regional level since some federal agencies have distinct regional planning authorities (e.g., Corps of Engineers, Bureau of Reclamation) that may involve drought plan development. It is imperative for district (regional) drought contingency plans developed by the Corps of Engineers, for example, to be compatible with state goals and policies.

Finally, regardless of the reasons for adopting formal drought response plans, states must expand the scope of their existing plans. It is essential for these plans to move beyond their initial impetus for plan development (i.e., reaction to crisis) and begin to address longer-term water-related issues as an integral part of the plan. This will require greater integration with other planning activities and improved coordination within and cooperation between levels of government.

Ironically, the policy themes of the past decade that we suggest may have influenced drought planning in some states in the United States are now being influenced by state drought planning efforts. This is evident in our suggestions for the need to (1) integrate drought planning with broader water management and planning efforts, (2) improve federal-state coordination in responding to drought, and (3) build on existing drought plans to address longer-term water issues in the context of federal-state relations. Because of the past decade's experience with state drought planning, issues involving the practice of federalism, water planning and jurisdiction authority, and appropriate responses to drought are becoming increasingly interrelated.

Internationally, interest in drought policy and planning has increased significantly within governments and international organizations in the past ten years and substantial progress is being made. The factors stimulating this interest are envisioned to be as complex as those discussed in this paper. Our discussion of factors affecting plan development in the United States should provide policy makers in these other settings with additional insights and understanding of this complicated but worthwhile process.

ABOUT THE AUTHORS

Donald A. Wilhite is professor of agricultural climatology in the Department of Agricultural Meteorology and director of the International Drought Information Center at the University of Nebraska-Lincoln. He specializes in studies of the impact of climate on society and societal response to climatic events, particularly drought. Dr. Wilhite is coeditor of *Planning for Drought: Toward a Reduction of Societal Vulnerability*, published by Westview Press in 1987. He has recently written a guidebook on drought preparedness for developing countries under sponsorship of the United Nations Environment Program. Dr. Wilhite is chair of the Committee on Applied Climatology of the American Meteorological Society.

Steven L. Rhodes is a political scientist with the Environmental and Societal Impacts Group at the National Center for Atmospheric Research (NCAR) in Boulder, Colorado. NCAR is sponsored by the National Science Foundation. Dr. Rhodes received his Ph.D. from the University of Colorado. His research interests include institutional responses to climate-related events and environmental policy in the context of climate change. He has written about such diverse topics as public policy and acid rain, renewable energy technology policy, the use of climate-related information for water resource management, desertification, and cleanup of contaminated federal facilities in the United States. From 1984 to 1989, Dr. Rhodes managed the preparation of the environmental impact studies for the new Denver International Airport project on behalf of the City and County of Denver.

REFERENCES

California Department of Water Resources. 1992. *The 1991 Drought Water Bank.* California Department of Water Resources, Sacramento, California.

GAO (General Accounting Office). 1979. *Federal Response to the 1976-77 Drought: What Should Be Done Next?* Washington, D.C., CED-79-26, January.

High Plains Associates. 1982. *Six-State High Plains-Ogallala Aquifer Regional Resources Study: Summary.* Report to the U.S. Department of Commerce and the High Plains Study Council, Washington, D.C.

Howitt, R.; N. Moore; and R. T. Smith. 1992. A retrospective on California's 1991 Emergency Drought Water Bank. Report prepared for the California Department of Water Resources. Sacramento, California.

Hy, R. J.; and W. L. Waugh, Jr. 1990. The function of emergency management. *In* W. L. Waugh, Jr., and R. J. Hy, eds. *Handbook of Emergency Management: Programs and Policies Dealing with Major Hazards and Disasters*; Chapter 2. Greenwood Press, Westport, Connecticut.

Karl, T. R. 1983. Some spatial characteristics of drought duration in the United States. *Journal of Climate and Applied Meteorology* 22:1356-66.

Lester, J. P. 1986. New Federalism and environmental policy. *Publius* 16(1):149-65.

NOAA (National Oceanic and Atmospheric Administration). 1982. *Climatic Impact Assessment*. Silver Spring, MD, EDIS-CEAS, January.

Palmer, W. C. 1965. *Meteorological Drought*. Research Paper No. 45, U.S. Weather Bureau, Washington, D.C.

Riebsame, W. E.; S. A. Changnon, Jr.; and T. R. Karl. 1990. *Drought and Natural Resources Management in the United States: Impacts and Implications of the 1987-89 Drought*. Westview Press, Boulder, Colorado, U.S.A.

Western States Water Council. 1987. *A Model for Western State Drought Response and Planning*. Western States Water Council, Midvale, Utah, U.S.A.

Wilhite, D. A.; N. J. Rosenberg; and M. H. Glantz. 1986. Improving federal response to drought. *Journal of Climate and Applied Meteorology* 25:332-42.

Wilhite, D. A. 1990. *Planning for Drought: A Process for State Government*. IDIC Technical Report Series 90-1, University of Nebraska-Lincoln.

Wilhite, D. A. 1991a. Drought Planning: A Process for State Government. *Water Resources Bulletin* 27(1):29-38.

Wilhite, D. A. 1991b. Drought planning and state government: Current status. *Bulletin of the American Meteorological Society* 72:1531-36.

Wilhite, D. A. 1992. *Drought Mitigation Technologies in the United States: With Recommendations for the Future*. Report of a Cooperative Agreement to the Soil Conservation Service, U.S. Department of Agriculture. IDIC Technical Series 92-1. University of Nebraska, Lincoln, Nebraska.

Chapter 14
Institutional Responses to the 1987-92 California Drought

James Rich

INTRODUCTION

From 1987 through water year 1992, California was hit by six consecutive years of dry weather. The Sacramento Basin Four River Index averaged about 10 million acre-feet (MAF) [12.33 million cubic decameters (dam^3)] for the period, compared to an average value of 18.4 MAF (22.7 $Mdam^3$). At least 400,000 acres (161,943 ha) of farmland were idled by the drought in 1991. Although the severity of the drought lessened in February and March 1992, significant idling of farmland is still expected for parts of California's fertile Central Valley (Rich, 1992a; California Department of Food and Agriculture, 1992).

RECENT AND EXPECTED INSTITUTIONAL RESPONSES TO THE DROUGHT

Institutional responses to the drought varied throughout the state. However, in only a few areas did the drought cause voters to decide on more water projects. Instead, some politicians and editorialists used the drought to promote reallocating water from agricultural to urban uses. The drought also gave impetus to various attempts to reform water use, such as increased regulation of groundwater. The creation of the state's Drought Water Bank (discussed later in this chapter) was the largest institutional response to the drought.

Because of the drought, Governor Pete Wilson directed his Office of Planning and Research and his Council on Growth to focus on water-related programs. The drought also gave impetus to the three-way dialogue between environmentalists and members of the urban and agricultural water communities. Their leaders now agree that the status quo is unacceptable, and they are attempting to find common ground and reach consensus. They support the recent signing of the "Best Management Practices" for urban water conservation, as well as state legislation (AB 3616), which requires efficient water management practices for agriculture. These measures will commit local urban and agricultural water districts to increasing conservation efforts and improving water management. Leaders from all three groups support groundwater banking, voluntary water transfers, the conjunctive use of groundwater and surface water, and the Drought Water Bank. One observer was struck by the "spirit of consensus" on water issues among competing sectors and interest groups in California. However, much work is needed to reach consensus on the overall solutions to California's water problems.

Some institutions are still changing in response to the drought. For instance, water recycling efforts are governed by state health regulations, which are being modified. If draft regulations are adopted, the number of approved uses for reclaimed waste water will rise from about 5 to more than 30. Those new uses would include irrigation of all food crops, plus residential landscapes, golf courses, and school yards (personal communication with Steve Kosower, California Department of Water Resources). John Fraser, of the Association of California Water Agencies, believes "the present [water] situation is intolerable." ACWA supports voluntary water transfers, water conservation, new water facilities, and waste water reclamation (Rich, 1992a).

The mission of the California Department of Water Resources (DWR) is to conserve, manage, and protect the state's water for beneficial uses. The drought intensified DWR's efforts to promote the conjunctive use of groundwater and surface water. For instance, DWR rehabilitated wells in a basin in the southern San Joaquin Valley that is recharged during wet periods. During the drought, the Kern County Water Agency, in cooperation with DWR, used the wells to supplement their reduced surface water supply. California is now negotiating with other states over an "interstate water bank" on the Colorado River System. Some see many possibilities in that area, which could increase supplies for urban Southern California during water-short years.

The drought caused some people to question California's water rights system, which limits flexibility in responding to water shortages. It also caused the U.S. Bureau of Reclamation (USBR) to give more attention to the efficient and environmentally sound management of existing water projects. For example, in 1991 and 1992, the USBR selectively released water from Shasta Reservoir and altered the operation of the Central Valley Project (CVP) to help spawning salmon (Rich, 1992a). Legislation passed in 1991 gives the USBR new power to move water where it is needed most during a drought (Anonymous, 1991).

The drought has focused increased attention on water transfers. That issue has divided the California water community and increased urban-agricultural tensions. John Fraser calls water transfers "a help, but not a cure-all" (Rich, 1992a). Several bills before the Legislature would facilitate water transfers in California.

Mike McGill, of the Bay Area Economic Forum, calls for "a market-based approach to California's water crisis." He believes we need a free market for water, with environmental safeguards and some government regulations. McGill notes that agriculture uses 80-85% of the state's developed water, yet directly contributes only $17 billion to a $750 billion economy (Rich, 1992a). Few parties in California want a completely free water market. Many want an open market, with some government regulation, to protect the environment and third parties. Many have called the Drought Water Bank a good first step toward such a water market.

CALIFORNIA'S DROUGHT WATER BANK

The biggest institutional change that occurred because of the California drought was the creation of the state's Drought Water Bank (DWB). Steve Macaulay, the engineer in charge of the DWB, believes it was a "success by almost every measure." It was recommended to Governor Wilson by his Drought Action Team and created on February 15, 1991 (Hoagland, 1991).

The DWB acquired water in three ways: by paying farmers to not irrigate their fields so the undiverted water would flow past their farms; by purchasing surplus water from local water districts; and by paying farmers or water districts to use groundwater instead of surface water (Rich, 1991a). The Bank set their buying price at $125/AF ($101/dam^3), based on analysis of crop budgets and crop water use estimates. Later they bought 20,000 AF (24,660 dam^3) of groundwater, at an average price of $40/AF ($32.44/dam^3). It cost about $5/AF ($4.06/dam^3) to administer the program. The water was sold at the state pumping plant in the Sacramento-San Joaquin River Delta for $175/AF ($142/dam^3) (Rich, 1992a; personal communication with Dave Marty, DWR, May 1992). Most of the price/cost difference was due to a 14% loss of "carriage water" in the Delta. (Carriage water is water lost in the transportation of DWB water to the state pumping plant.)

About 120,000 AF (147,960 dam^3) were bought by the DWB directly from farmers. The Bank sold about 390,000 AF (480,870 dam^3), mainly to urban water districts. The rest of the 820,000 AF (1,011,000 dam^3) of DWB water went to the State Water Project for future use or was used as carriage water. (The State Water Project is a series of reservoirs and canals that take California water from the north to the south). The Bank also provided 50,000 AF (61,650 dam^3) to wildlife refuges (Rich, 1992a; personal communication with Dave Marty, DWR, May 1992). Refer to Table 1 for more information.

Macaulay admits that the program had a few problems (Rich, 1992a). For instance, the $175/AF ($142/dam^3) cost at the Delta meant that some farmers who needed DWB water to save a crop, trees, or vines did not get any because they could not afford it. It was also difficult to identify and compensate third parties hurt by negative impacts of sales to the Bank. For example, Yolo County asked that the DWB provide $130,000 to compensate them for the benefits they claimed were paid to farm workers idled by sales to the DWB. This issue is not yet resolved. Some counties, wary of these negative impacts, have placed certain restrictions on the sales. Yolo and Butte counties placed a 2% levy on sales to the DWB to fund groundwater-level studies. Tehama County

Water Source	Acre-Feet (rounded)
Denial of Irrigation by Farmers	420,000
Groundwater Exchanges	260,000
Surface Water from Reservoirs	140,000
Total	**820,000**
Delta Carriage Water Requirements, Technical Corrections	-160,000
Net Supplies	660,000
ALLOCATIONS	
Urban Uses	307,000
Agricultural Uses	83,000
State Water Project	270,000

Table 1. 1991 Drought Water Bank purchases and allocations (California Department of Water Resources, 1992a).

recently passed a law to prevent almost any of their groundwater from being used outside the county (Anonymous, 1992).

Macaulay believes concern over winter-run salmon may affect 1992 Bank operations. The 1992 DWB will be smaller than the 1991 Bank. The Bank will probably offer only \$50/AF (\$40.55/dam^3) for water. Most of the water would come from surplus reservoir storage or groundwater exchanges. More of the 1992 Bank's water is expected to be sold to agriculture (Rich, 1992a).

According to Dr. Richard Howitt, University of California-Davis, the success of the DWB demonstrates that during a drought, a significant number of farmers will sell their water. However, even with serious surface water shortages in some areas, cities did not buy up all the water available. Also, the unsold Bank water showed that many cities did not want to meet their unfilled water needs at any price. The DWB showed that the water market has a supply function, a demand function, and a market-clearing price. As Howitt puts it, water in California is changing from a resource controlled by politics to one controlled by an economic market (Rich, 1992a). Howitt believes that with the exception of sugar beets, all the major crops idled by sales to the DWB were low labor input crops. Few, if any, regions had DWB-induced acreage declines that were greater than normal yearly fluctuations.

The 1991 water transfers in California constituted a controlled, regulated water market. Environmental and water agency concerns, and third-party impacts, will prevent the formation of a completely free water market. Because of the drought, we should have more conjunctive use of groundwater and more water sales involving groundwater. Howitt foresees flexible Water Bank pricing, plus spot and futures markets for water, and even options contracts for water (Rich, 1992a). He sees more economic and fewer political coalitions controlling water. He believes the 1991 DWB showed that California can develop up to 800,000 AF (986,400 dam^3) of water in a short time through water transfers. The Water Bank may become permanent, with a role for private water transfers as well.

Many farm leaders believe that water transfers through the Water Bank were beneficial to farmers. They expect the 1992 bank to make more affordable water available to agriculture. However, some farm leaders still fear that the Bank is an attempt to get the farmers' water rights (Rich, 1992a). Some farm groups and local governments worry about the third-party impacts of farm water sales on rural communities. They ask, "Do payments made to farmers selling their water trickle down to those who lose their jobs or businesses because the farmer idles his land?" Also, is it easier to avoid local economic impacts (through localized acreage caps, perhaps) than to mitigate them once they've occurred?

The drought also affected Governor Wilson's new long-term state water policy, announced in April 1992 (California Governor's Office, 1992). In that statement, the governor advocated "sustained-yield ground water management," a controlled water market, a state takeover of the federal CVP, and the construction of more off-stream storage facilities.

MITIGATION OF DROUGHT IMPACTS

Because of the changes and responses of the many institutions involved with water in California, the economic and social impacts of the drought were lessened. For instance, agricultural impacts were not as great as many had expected, given the severity of the drought. According to one estimate, agricultural water use in 1991 declined only 7%, with a loss in gross farm revenues of less than $600 million. The main reason for this relatively mild impact was a 104% increase in groundwater use (compared with predrought levels) during 1991. This was made possible, in part, by 1,300 new agricultural wells (Rich, 1992a). The drought also increased the reluctance of many lenders to make farm loans or credit available to farmers without a secure, dependable water supply.

The drought's heaviest impact may have been on fish and wildlife and their habitats. For instance, 1990 was the first year that water contractors experienced significant drought impacts, while some aquatic environments were stressed earlier in the drought. In early 1992, federal agencies made decisions to reduce water deliveries to many farmers, to make more water available for environmental purposes (Rich, 1992a).

Thanks to the creative short-term solutions of urban water managers, urban impacts of the drought were minimized. For example, the San Francisco Water Department (SFWD) serves a population of 2.6 million. About 85% of their water usually comes from the Sierras, and SFWD has only a limited ability to augment that water. In 1991, they purchased 65,000 AF (80,140 dam^3) of supplemental water from the DWB, Placer County, and Modesto Irrigation District (Rich, 1992a). Those purchases enabled the Department to institute a 25% rationing program instead of a planned 45% rationing regime without the supplemental supplies.

SFWD had a mandatory 25% rationing plan for three of the five years from 1987 to 1991. Their goal was a 10% drop in indoor water use, and a 60% reduction in outdoor use. In 1991, they had a 30% reduction in overall water use. In response to the drought, the Metropolitan Water District of Southern California purchased additional Colorado River water, instituted a 20% rationing program, and modified its distribution rules to favor farmers (Rich, 1992a).

Water districts have helped each other survive the drought. For instance, Yuba County Water Agency's (YCWA) 966,000 AF (1,191,000 dam^3) Bullards Bar Reservoir has had surplus water available for drought relief since 1987. YCWA has worked with government agencies to maximize the environmental benefits of their water releases. In 1991, YCWA sold 157,200 AF (193,830 dam^3) to the Water Bank, 28,000 AF (34,520 dam^3) of which went to the Department of Fish and Game for wildlife use. However, YCWA now fears that "the State appears to be targeting us for reduction of our water rights, due to our water sales" (Rich, 1992a). Unless they receive more protection for those rights, they might not make any more water sales.

Gary Bucher, a Kern County Water Agency (KCWA) manager, believes that supplemental surface water, increased groundwater pumping, and east-to-west water transfers allowed almost all their permanent crop acres to record near-normal yields in 1991. KCWA helped arrange some complicated and creative intracounty water transfers. Often those deals involved water exchanges with DWR or the use of DWR facilities. Most of KCWA's supplemental water cost them \$100 or \$140/AF (\$81 or \$114/dam^3). The delivered cost of DWB water was about \$200/AF (\$162/dam^3). KCWA would have bought more DWB water if it had cost \$125/AF (\$101/dam^3) or less (Rich, 1992a).

Bucher estimates that water shortages idled 130,000 KCWA acres (52,630 ha) in 1991 (Rich, 1992a). Yet in July, no one could find evidence of any drought-induced job loss. One reason for this may be that about half of the lost KCWA acres were made up for by increased plantings on the east side of the county, which had adequate groundwater (Rich, 1991b).

There were some problems with local water transfers in Kern County. Some farmers wanted to pump and export groundwater from an overdrafted basin. Some east-side farmers wanted west-side farmers to pay them to fallow their land, and then pay

KCWA to transport the purchased water to the west side. But the biggest problems with most of the proposed local water transfers were environmental. For example, KCWA was unable to take any water from the Kern Water Bank during 1991, because they could not get a permit in time to allow them to turn on their pumps in 1991. This was due to state environmental laws designed to protect endangered species (Rich, 1992a).

Bucher believes that despite the severity of the drought, KCWA was able to prevent an economic disaster in 1991. However, if water shortages prevent many of KCWA's growers from planting crops again in 1992, he fears some will go bankrupt (Rich, 1992a).

OTHER DROUGHT IMPACTS

California's electric utilities responded to the drought-induced loss of hydropower generation by substituting more expensive fossil fuels. By one estimate, this cost ratepayers $2.93 billion from 1987 through 1991 (California Department of Water Resources, 1991).

The only sector of the urban economy that was significantly affected by the drought was the "green industry": firms or individual persons who design, create, install, or maintain urban landscapes. A preliminary estimate found $460 million in 1991 drought-induced direct losses of equivalent gross revenue to the California green industry, compared to about $7 billion in equivalent gross revenue in 1990[1]. In response to this, the industry lobbied to oppose local moratoriums on landscape installation, local water-use quotas, and ordinances mandating water-conserving landscapes (Rich, 1992b). Some industry leaders fear the drought has permanently hurt their industry.

According to one estimate, the drought imposed a direct economic cost on California agriculture of about $450 million in 1990. The cost was attributed to reduced yields, increased farm and ranch costs, and lost output from about 190,000 drought-idled acres (76,900 ha) (California Department of Water Resources, 1991). Another preliminary estimate indicated that the continuing drought resulted in a net idling of about 450,000 acres (182,200 ha) in 1991 (Rich, 1991b). Lost gross farm revenue from those acres was about $300 million. About half of the idled acres would have been planted in wheat or cotton. Most of the idled acres were on the west side of the southern San Joaquin Valley.

Despite the drought, cash receipts from farm marketings in California reached a record $18.3 billion in 1990, up from $17.5 billion in 1989 (California Department of Food and Agriculture, 1992). However, those receipts declined slightly in 1991, partly because of the greater number of drought-idled acres. There is some concern that some

[1]Some sectors of the green industry, such as public parks and freeway right-of-ways, have no revenues. In these cases, maintenance budgets and payrolls were used as proxies for the value society places on keeping them green.

acres of permanent crops, stressed by suboptimal irrigations in 1991, will produce less in 1992 and 1993. On the other hand, official statistics indicate that the drought had a negligible effect on agricultural unemployment (Rich, 1991b).

CONCLUSION

On April 6, 1992, Governor Wilson proposed a comprehensive new water policy. If adopted, it would help lessen the impacts of future droughts. The governor's key proposals include:

- Restoring the Delta and improving water supply through the linked processes of completing South Delta Facilities and implementing revised Delta standards for the protection of fish and wildlife.
- Building additional offstream storage facilities such as Los Banos Grandes, Los Vaqueros, and Domenigoni Reservoirs to meet future water requirements.
- Encouraging water marketing, a key element in the water policy that requires careful consideration by future legislation so that these criteria are met: Water transfers must be voluntary and protect the water rights of sellers. Water transfers must not harm fish and wildlife resources or their habitats. There must be assurance that transfers will not cause groundwater overdraft or degradation of groundwater basins. The transferred water must be used efficiently. Water districts and agencies that hold water rights or contracts to transferred water must have a strong role in determining what is done.
- Promoting water conservation, which will continue to play a vital role in reducing overall demand. The governor looks to both urban and agricultural users to share the burden and benefits of conservation.
- Addressing the problem of groundwater overdraft. The governor is urging state and local entities to develop long-term plans to recharge aquifers during times of moderate and heavy runoff, and to tap this storage when required during dry years (California Department of Water Resources, 1992b).

In conclusion, institutional responses to the 1987-92 drought helped to lessen drought impacts. For instance, California agriculture did well, despite the severity of the recent drought, because of effective responses by the agricultural and water communities. Farmers, ranchers, farm workers, agribusiness leaders, farm advisors, water district managers, and local, state, and federal government officials worked together to devise innovative solutions to the serious problems caused by the drought. A record number of wells were drilled or deepened. A successful water bank helped assure normal yields on 113,000 acres (45,750 ha) of once-endangered permanent cropland in the San Joaquin Valley. Farmers used computers and localized, farmer-oriented weather data, in conjunction with new irrigation management techniques, to significantly reduce water usage. Cropping patterns were changed to produce more revenue with less water. And innovative water sales and exchanges were made, thanks to a spirit of cooperation that existed throughout the agricultural and water communities.

ABOUT THE AUTHOR

James Rich has worked as a government economist since 1976. He holds two bachelor degrees from the University of Washington and an MBA in applied economics from the University of California-Berkeley. He has been an economist with the California Department of Water Resources since 1980, specializing in agricultural economics. He has forecasted the demand for irrigation water, estimated the price elasticity of demand for water, done cost/benefit analysis of water projects, and estimated the economic impacts of water marketing and the 1987-92 California drought.

REFERENCES

Anonymous. 1991. U. S. Senate Energy Panel passes drought legislation. *Los Angeles Times*, 27 September.

Anonymous. 1992. Butte area water sales debated. *Marysville Appeal-Democrat*, 11 January.

California Department of Food and Agriculture. 1992. Press release (#92-24), 30 January. California Department of Food and Agriculture, Sacramento.

California Department of Water Resources. 1991. *California's Continuing Drought, 1987-91*; pp. 19-23, 36. California Department of Water Resources, Sacramento.

California Department of Water Resources. 1992a. DWR fact sheet on water bank operations (provided by Dave Marty). California Department of Water Resources, Sacramento.

California Department of Water Resources. 1992b. *Report to the California Water Commission*; p. 4. California Department of Water Resources, Sacramento, 1 May.

California Governor's Office. 1992. Press release by Office of California Governor Pete Wilson, 6 April. California Governor's Office, Sacramento.

Hoagland, R. 1991. Third-party impacts of State Water Bank purchases. DWR memo to Steve Macaulay, 26 June. California Department of Water Resources, Sacramento.

Rich, J. W. 1991a. Draft material—Impact of the State Water Bank on California agriculture, 30 October. California Department of Water Resources, Sacramento.

Rich, J. W. 1991b. The impact of the 1991 drought on California agriculture. DWR memo to Ray Hoagland, 20 November. California Department of Water Resources, Sacramento.

Rich, J. W. 1992a. CII meeting. DWR memo to Ray Hoagland, 23 March. California Department of Water Resources, Sacramento.

Rich, J. W. 1992b. Impact of the drought on the green industry. DWR memo to Ray Hoagland, 3 January. California Department of Water Resources, Sacramento.

Chapter 15
Drought in China: Present Impacts and Future Needs

Kerang Li and Xianchao Lin

INTRODUCTION

In recent years, frequent and severe drought has become a major climate disaster throughout the world, adversely affecting developed and developing regions as well as the environment. Drought can occur in virtually all climate regimes, but it is especially damaging in monsoon, semiarid, and arid regions. The frequency, duration, and spatial extent of drought and the magnitude of agricultural losses and potential environmental effects make it China's greatest natural disaster. Recent droughts have attracted attention to its impacts and the need for improved assessment and management procedures to help mitigate its adverse effects.

How can society reduce the impacts of drought? Although many tactics can be employed to mitigate the effects of drought, an important first step is the establishment of adequate monitoring programs to provide early warning as drought conditions emerge. Human activities may also contribute to the occurrence and severity of drought. Overgrazing, poor cropping methods, deforestation, overpumping of groundwater, and improper soil conservation techniques often exacerbate the magnitude of impacts. However, through the adoption of proactive strategies, losses and hardships in China can be reduced.

REGIONAL DISTRIBUTION OF DROUGHT IN CHINA

The occurrence of meteorological drought in China was evaluated on the basis of departure-from-normal precipitation and the duration of the event. Two categories of drought were defined, moderate and severe, on the basis of the work of Feng et al. (1985).

Drought Duration	Moderate Drought (%)	Severe Drought (%)
> 3 months	-25 ~ -50	Over -50
2 months	-50 ~ -80	Over -80
1 month	Over -80	

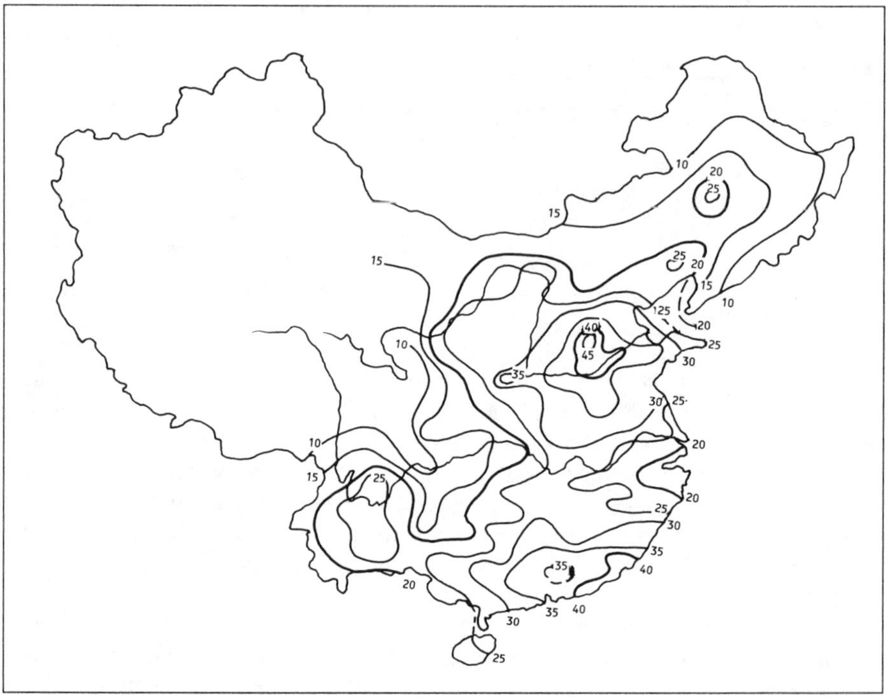

Figure 1. The distribution of the number of drought occurrences in China, 1951-90.

Figure 1 gives the distribution of the number of drought occurrences in China, 1951-90. Two regions, northwestern China (arid year-round) and Qinghai-Xizhang Plateau (cold high plateau region), were not included. Based on the number of drought occurrences and an analysis of the average number of drought months (not shown), regional differences in drought occurrence are notable. The five main drought-prone regions in China are discussed below (see Figure 2).

The Huang-Huai-Hai Plain Region

This region covers the middle and lower reaches of the Yellow River, the valleys of the Huaihe River, and the Haihe River (i.e., the North China Plain, Loess Plateau, and Inner Mongolia Plateau). It is located in the semiarid and semihumid monsoon region and is characterized by high annual precipitation variability and uneven seasonal distribution of rainfall. This region of China is the most frequently affected by drought;

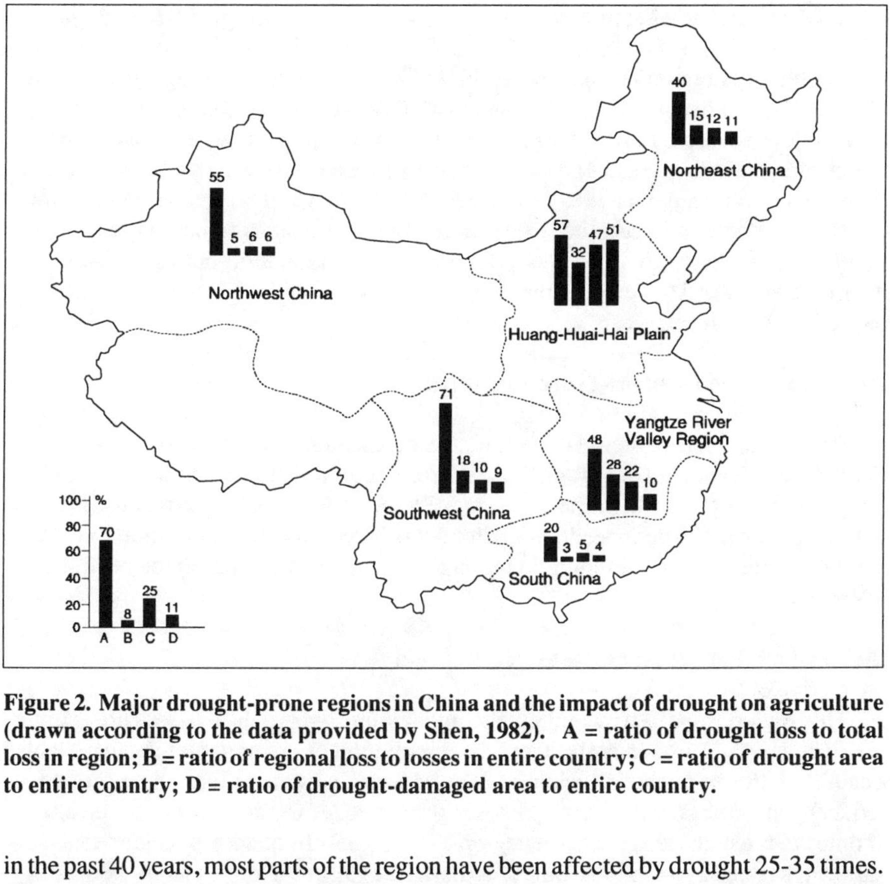

Figure 2. Major drought-prone regions in China and the impact of drought on agriculture (drawn according to the data provided by Shen, 1982). A = ratio of drought loss to total loss in region; B = ratio of regional loss to losses in entire country; C = ratio of drought area to entire country; D = ratio of drought-damaged area to entire country.

in the past 40 years, most parts of the region have been affected by drought 25-35 times. Also, drought in this area is more intense, of longer duration, and greater in spatial extent. The North China Plain has been affected more than 35 times, or nearly every year during the 1951-90 period. The greatest number of drought occurrences (up to 45 times) in the country occurred in the southern part of Hebei Province. Droughts also tend to be of longer duration (3-4 months) here, as opposed to 2-3 months in other parts of the region. In the past 40 years in this region, severe droughts were observed in 13 years (1955, 1957, 1960, 1962, 1965, 1968, 1972, 1978, 1981, 1983, 1986, 1987, and 1989).

South China and the Middle and Lower Reaches of the Yangtze River Region

Though this region is situated in a humid region of China, drought occurs with relatively high frequency. This is because of the large annual variability of precipitation and its uneven seasonal distribution. The region is the second-largest drought-prone region, after the Huang-Huai-Hai Plain drought-prone region. In the past 40 years, most of the areas in the region have experienced 20-35 droughts. The greatest number, 40-43, was recorded in the coastal zones of Guangdong and Fujian Province, where drought has been recorded nearly every year. Average drought duration in this region is 2-3 months each year. However, in the coastal zone of Guangdong Province, this figure reaches 3.3 months per year.

Southwest China Drought-Prone Region

Located in southwestern China (i.e., the northern and central parts of Yongnan Province and Xichang District of Sichuan Province), this is the smallest of the five drought-prone regions. In the past 40 years, 20-25 droughts have been recorded. Parts of this region have experienced 26-30 droughts. Average drought duration is 2.0-2.5 months per year; the longest duration droughts (2.8 months) occur in the central part of Yongnan Province.

The Western Part of Northeast China Region

This region is located in a relatively high latitude, where the temperature is quite low. The growing season is short and the distribution of precipitation is fairly stable because of the frequent intrusion of low pressure systems. Drought occurrence frequency in this region is relatively low. In the past 40 years, drought has occurred 10-20 times in most of the areas of the region. The highest frequency of occurrence, 22-26 times, has been recorded in Baicheng District of Jilin Province and in the western part of Liaoning Province. The average drought occurs once every 1.5-2.0 years. For the region as a whole, the average annual drought duration is 0.5-1.5 months per year, but in Baicheng District the average is 2.1 months.

Northwest China Region

Northwestern China is an arid region characterized by year-round water deficit, extremely high wind erosion, and, consequently, very low agricultural productivity. It represents one of the most arid areas in China, with an average annual precipitation of less than 250 mm. Precipitation gradually decreases from southeast to northwest in the region. Annual rainfall is less than 50 mm in the eastern Tarim Basin and western

Qaidam Basin. In the Turpan Basin, the 10-year mean annual rainfall from 1961 to 1970 was 3.9 mm, which is the record minimum rainfall in China. Drought duration increases gradually from northeast to southwest and from east to west across the region. On average, southern Xinjiang has the longest mean duration of drought, more than 250 days per year. Qinghai and northern Gansu Province have the second-longest mean duration of drought.

For China as a whole, the average occurrence of severe and nationwide drought has been 7.5 times each year from 1951 to 1990. Floods, the second most frequent natural hazard, occurred 5.7 times. Regional differences in drought distribution are apparent. Generally speaking, droughts occur more frequently in northern China than in southern China, and they usually occur with greater intensity.

SEASONAL CHARACTERISTICS OF DROUGHT IN CHINA

China is situated in a continental monsoon region. Its annual and regional distribution of precipitation are quite uneven. Therefore, seasonal occurrences of drought differ greatly from one region to another (see Figure 2); agricultural impacts also differ significantly. In general, drought may occur in any season of the year. In China, droughts are usually classified according to their season of occurrence (i.e., spring drought, summer drought [including early summer drought and midsummer drought], winter drought, or season-persistent drought).

Huang-Huai-Hai Plain Region

Droughts that occur in this region usually do so during the crop-growing period, from March to October. Thus, droughts are associated primarily with the spring, summer, or autumn period. Spring droughts occur most frequently, followed by summer drought and summer-autumn persistent drought; spring-summer persistent drought occurs with the lowest frequency. In the past 40 years, spring drought occurred 34 times. Two Chinese sayings describe such a situation as "nine years out of ten will be hit by drought" and "rainwater in spring is as valuable as oil." Spring drought usually begins in March and extends to May or June.

The Yangtze River Valley Region

Droughts in this region occur mainly in summer and extend into the autumn months. Occurrence is usually from July to September. Thirty-four of the last forty years were affected by summer drought. In some years, summer droughts may even extend to October or November. In that case, they turn into summer-autumn persistent drought.

South China Region

The climate in this region is characterized by warm temperatures and abundant precipitation. The region is characterized by a year-round growing season, and droughts may occur at any time of the year, although winter-spring droughts (droughts that begin in late autumn and continue into early spring) are the most common type, followed by autumn droughts.

Southwest China Region

Droughts in this region are usually of relatively limited spatial extent and occur mainly during the winter and spring months in the form of winter-spring persistent drought.

Northeast China Region

Droughts in this region occur mainly in spring and summer months; spring droughts are the most frequent. Spring droughts were recorded 25 times and summer droughts (i.e., July or August) 21 times in the past 40 years.

In summary, seasonal differences in drought occurrence in China are quite obvious. Generally, during the spring months, northern China is dry while southern China is vulnerable to floods. During the midsummer months, the situation is reversed. Thus, droughts are more common in northern China during the spring; southern China records most of its droughts in midsummer. For China as a whole, winter-spring droughts or spring droughts occur most frequently, are most severe, and are of longest duration. Statistically, the occurrence frequency of these droughts is more than 40%. It is important to note that all the severe drought years in China are associated with winter-spring persistent droughts. Summer droughts also occur frequently in China. In particular, weather conditions in the middle and lower reaches of the Yangtze River are dominated by the western Pacific subtropical high.

DROUGHT OCCURRENCE IN CHINA IN THE PAST 500 YEARS

An index (I_D) was used to calculate the relative dryness/wetness for each year for the period 1470-1977 (508 years) from data for 100 stations, using the equation shown below.

$$I_D = 2\frac{D}{N}$$

where D = the total number of stations where drought and severe drought were recorded in a given year; N = the total number of stations. I_D then expresses relative dryness or

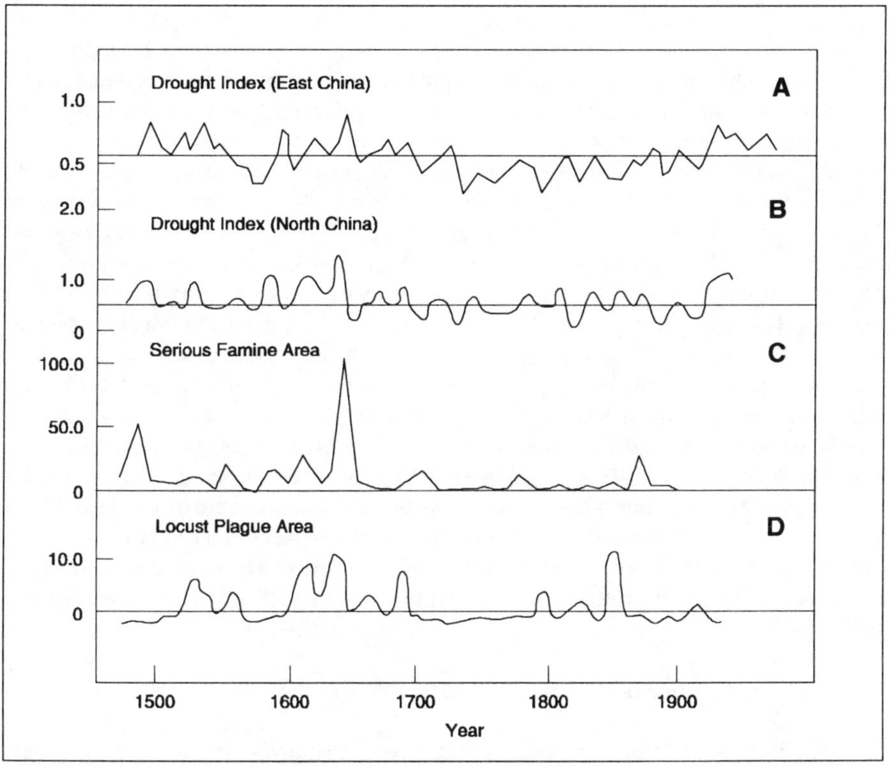

Figure 3. I_D of China in the past 500 years and the number of famine-stricken counties and locust disaster-stricken counties.

wetness for each year as the ratio of drought and severe drought stations to the percentage of the total stations, times two. This index reflects the area affected by drought on a scale ranging from 0.0 to 2.0, as shown in Figures 3a and 3b for eastern and northern China, respectively (Zhang, 1981). This analysis indicated that during the past 500 years, the 50-year record severe drought was observed 18 times. Droughts of this magnitude occurred mainly in the sixteenth and seventeenth centuries and in the 1920s, 1930s, 1960s, and 1970s. The severe drought years were 1528, 1640, 1641, 1721, 1785, and 1972.

The curves of 10-year running mean values of I_D in east China, 1470-1977, and in the North China Plain region, 1470-1949 (see Figures 3a and 3b; Chen, 1986), show that in these two regions, three distinct periods can be identified in the past 500 years.

The first one, lasting 213 years, was a dry period during which 10 severe drought years were identified. The second period was wet, lasting 199 years, during which only 4 severe drought years were recorded. Since 1891, I_D has remained higher, which indicates that a new dry period has developed. Until recently, only 4 drought years had occurred. If a 200-year period is considered the average persistent period for the alternation of dry and wet periods, the present dry period can be expected to continue. The results above show that in the past 500 years, the sixteenth and seventeenth centuries turned out to be dry while the eighteenth and nineteenth centuries were wet. In general, conditions during the twentieth century have been dry. It is interesting to note that during each period, a number of shift episodes of dryness and wetness with time scales of 10-20 years have existed. The two most significant decade-long dry episodes were from 1581 to 1590 and from 1634 to 1643. In particular, in the latter episode in the eastern part of China, the number of stations where drought and severe drought were recorded reached almost 50% of the total number of stations. The same episode in northern China appeared to be even worse, with most of the areas affected by persistent drought, except over small portions of the region. This was considered the worst drought in the past 500 years. Further analysis shows that the 7-year severe drought (1637-43) in northern China originated in the northern part of Shanxi Province in 1627 and ended in Changsha, Hunan Province, in 1646, a total of 20 years. The economy in nearly 15 provinces in the Yellow River Valley and the Yangtze River Valley was seriously affected.

Drought Occurrence in the Past 40 Years

The patterns of dryness and wetness displayed over the past 500 years have also been a climatic feature of the past 40 years. As a whole, China was relatively wet throughout the 1950s and rather dry in the 1960s and 1970s. In the 1980s, the Yangtze River Valley and Huai River Valley were wet while the northern and southern parts of China were dry. Present conditions are relatively wet, but the region is still within a dry period. For China as a whole, the most severe and longest duration droughts observed in the past 40 years occurred in 1955, 1959, 1960, 1962, 1963, 1965, 1966, 1971, 1972, 1978, 1979, 1986, and 1988, a total of 13 years. Most of these droughts occurred in the 1960s and 1970s. Although droughts can last for 5-7 months or longer in some locations, they usually occur from March to October and last for 3-4 months.

In summary, the persistent phased quasi-periodic characteristics of drought in China vary with time. This persistent character is illustrated by the seasonal persistence of dry conditions. For instance, most severe droughts are characterized by spring-summer persistence, at times for several consecutive years. As mentioned above, the severe drought that occurred from 1627 to 1646 was unusual in the history of the past 1,000 years, in terms of spatial extent, duration, and intensity. However, similar examples can also be identified in the record of the most recent 40-year period. For example, the middle and lower reaches of the Yangtze River were affected by drought for 4 consecutive years from 1958 to 1961 and again between 1966 and 1968. Furthermore, the northern region of China suffered continuous drought throughout the 1980s. Severe droughts usually occur over relatively large areas.

IMPACTS OF DROUGHT ON AGRICULTURE

China is an agricultural country. The importance of drought lies in its impacts on agriculture, specifically on crops, trees, vegetation, and grassland. The indirect effects of drought often extend well beyond the reduction of agricultural output, at times leading to famine, the spread of refugees, social instability, and even death.

Drought-Affected Areas and Drought-Damaged Areas

In China, droughts are generally considered to affect extensive areas while floods usually inundate narrow or localized belt-type areas. According to the statistics from 1950 to 1989, the annual average area of cultivated land affected by drought was 20.677 million ha. This accounts for 60% of the total cultivated land affected by climatic disasters and 17.2% of the total area cultivated for grain. Among the drought-affected areas, the drought-damaged area (defined as areas with a total grain yield reduction of more than 30%) averaged 8.004 million ha. This accounts for 6.7% of the total area cultivated for grain. The drought-affected and drought-damaged areas during the severe drought years of 1959, 1960, 1961, 1972, 1978, 1986, and 1988 are given in Table 1.

Based on statistics from 1950 to 1979, climate disasters in China resulted in a total grain reduction of about 3,062.2 x 10^8 kg, or 4.8% of the total grain output in China

Drought Years	% Area Drought-Affected	% Area Drought-Damaged
1959	33.8	11.2
1960	38.0	16.2
1961	37.9	18.7
1972	30.7	13.6
1978	40.2	18.0
1986	31.1	14.8
1988	32.9	15.3

Table 1. Percent of cultivated area affected and damaged by drought, 1950-89.

Type	Total Reduction ($x10^8$kg)	Percent
Drought	1,532.1	50.0
Flood	844.3	27.6
Drought and Flood	540.0	17.6
Freezing	145.9	4.8
TOTAL	**3,062.3**	**100.0**

Table 2. Total reduction of grain yield caused by various climate disasters.

during the period. The annual average loss of grain was 102 x 10^8 kg. Table 2 gives the reduction of grain yield, by climatic disasters. It is obvious that the greatest reduction was caused by drought. The total loss due to drought was as high as 1,532 x 10^8 kg, or 50% of the total grain loss (Shen, 1982).

Regional Distribution of Drought-Affected and Drought-Damaged Areas

Agricultural impacts differ significantly from region to region, reflecting the regional distribution of drought-affected and drought-damaged areas (see Figure 2). Figure 2 shows that the Huang-Huai-Hai Plain region is identified as the most drought-affected area; impacts are also greatest in this region. This region accounts for 46.5% of the total affected area in the country and 50% of the drought-damaged area. Grain production losses because of drought account for 57.4% of the total loss in the region and 32.1% of the total grain loss in the country. In other regions, such as southwestern China, northwestern China, the middle and lower reaches of the Yangtze River, and northeastern China, the loss of grain output in each region due to drought also accounts for 50% or more of the total loss caused by various disasters in each region (see Figure 2 and Li and Lin, 1991).

Historical Variation of Drought-Affected and Drought-Damaged Areas

The periodicity of drought-affected and drought-damaged areas in China has been significant in the last 40 years. During this time, three wet periods and three dry periods occurred. The periods of 1958-62, 1971-81, and 1986-88 were observed as significant drought periods (i.e., more than 20 million ha were affected). During the periods 1950-57, 1963-70, and 1981-85, less than 20 million ha were affected. In total, seven severe drought years (1959, 1960, 1961, 1972, 1978, 1986, and 1988) were recorded during the significant drought periods. In each instance, the drought-affected areas encompassed more than 30 million ha.

The variability in the occurrence of drought-affected and damaged areas seems to have increased during the last 40 years. Averaged by decades, the drought-affected areas in China from the 1950s to 1980s are 12.6, 21.7, 26.1, and 24.1 million ha, respectively. Although the area affected in the 1980s was slightly less than that in the 1970s, it still represents a considerable increase over the 1960s and 1950s. This trend is consistent with the general trend of summer aridization in China (Song, 1991).

Based on precipitation anomalies for 160 stations in the most recent 40-year period (Li and Lin, 1992), precipitation was above normal during the 1950s and below normal during the 1960s and 1980s. In particular, precipitation north of the Yangtze River, such as in the southern part of northeastern China, Inner Mongolia, northern China, Loess Plateau, Shandong Province, and Huaihe River Valley, was above normal during the 1950s and 1960s. In the 1970s and 1980s, precipitation was below normal. In northwestern China, precipitation was above normal during the 1950s-1970s and below normal during the 1980s.

Drought-affected areas have increased dramatically in the last 30 years in the Huang-Huai-Hai Plain region, northern China, and northwestern China. The drought-affected areas in southwestern China have also increased, but by a lesser amount. By contrast, drought in the middle and lower reaches of the Yangtze River and in southern China has shown a significant reduction from the levels of the 1950s and 1970s.

Relationship of Drought, Famine, and Locust Disaster

The shortage of food and herbage caused by drought often results in famine, pestilence, locust disaster, and death in countries and regions that are less economically developed. Although based on incomplete data (Chen, 1986), research shows that drought-related death tolls of more than 1,000 persons were recorded 24 times in China for the period from 180 B.C. to 1949 A.D. The total death toll during this period was 11.854 million.

Figures 3a and 3b depict the 10-year moving average of the drought index (I_D) in eastern and northern China, respectively. This figure indicates the size of the areas affected by drought. Figures 3c and 3d represent the 10-year moving averages for the number of famine-stricken and locust-stricken counties, respectively. The number of counties seriously affected by famine is in general agreement with the variation of I_D. The positive relation between famine and drought is significant: the correlation coefficient is 0.413 and the level of significance is $\alpha = 0.01$. Figure 3a shows that the eastern part of China was arid from 1479 to 1691 and in the post-1891 period, but wet from 1692 to 1890. The situation was similar in northern China. During the arid period, 1479-1691, China experienced its most severe arid stage of the last 500 years, extending from 1637 to 1643. The area affected and the duration was also the greatest on record. This period also records the largest number of counties experiencing severe famine.

Another severe drought appeared in China from 1875 to 1895, during a climatically humid period. The most severely affected areas were in northern China, the Loess Plateau, and the middle and lower reaches of the Yangtze River. Severe famine was recorded as a consequence of the drought. Drought conditions can also result in the spread of locusts (Figures 3b and 3d). The correlation coefficient is R=0.423, the level of significance 0.01. This clearly indicates that drought provides favorable climate conditions for the occurrence and development of locust disasters.

POTENTIAL DROUGHT IMPACTS ON THE NATURAL ENVIRONMENT

In addition to having impacts on agricultural output, economy, and society, drought also influences water quantity and quality and land degradation processes (e.g., desertification). Environmental degradation may seriously threaten the habitability of China, particularly in the north.

Persistent Reduction of Water Resources

Although China is rich in the total volume of water resources available, the nation's per capita availability is quite low. Average annual precipitation in China is 648 mm, slightly lower than the world average of 800 mm. The average annual surface runoff volume in China is 2,711.5 billion m^3, the sixth highest value in the world. However, the unit area runoff yield is only 284 mm, less than any other region except the African and Antarctic continents. Per capita water availability is only 2,630 m^3 in China, about one-fourth of the world average and one-fifth of the average for the United States. At present, the annual water deficiency for China is 50 billion m^3. It is predicted that at current population growth rates and with the decrease in cultivated land areas, the per capita water availability will decrease further (Ministry of Water Conservancy, Hydrological Department, 1987).

Reduction of Surface Runoff and Total Amount of Water Resources in Northern China. As a result of drought, water shortages have occurred widely in northern China in recent years. For example, the water resources of northern China since 1955 have decreased at a rate of 9.3 billion m^3 per decade. In the severe drought years of 1965, 1968, and 1972, the total volume of water resources in north China was only 30.0 billion m^3 (Song, 1991). Table 3 gives the average values of surface runoff in each decade in the Yellow River Valley of northern China, 1956-84, and their ratio with normals. This table shows that the runoff volume decreased from the 1950s to the 1980s. Runoff volume decreased most significantly in the lower reaches of the Yellow River, from 142 million m^3 in the 1950s to 54 million m^3 in the 1980s.

In northwestern China, a gradual decrease in runoff volume was recorded in the past few decades. The decrease is demonstrated by the abundant runoff volume in the

Region	1956-84 Mean	1956-59 Mean/Ratio	1960-69 Mean/Ratio	1970-79 Mean/Ratio	1980-84 Mean/Ratio
Yellow River Valley	6.72	2.61/1.26	7.73/1.12	5.80/0.86	5.36/0.79
Lower Yellow River	8.30	11.78/1.42	9.01/1.09	8.09/0.93	4.49/0.54

Table 3. Mean runoff (billions m³) in each decade and their ratios with normals, 1956-84.

1950s, volume decline in the 1960s, and decrease of runoff after the 1970s. The decrease is more significant after the runoff enters the plain areas and the impacts of human activities are introduced. For instance, in Xinjiang Province the total annual runoff volume was reduced from 80.51 billion m^3 in the 1960s to 72.45 billion m^3 in the 1970s, a reduction of about 10% (Li and Lin, 1991).

Changes in Surface Water Supplies. The total area of lakes in China is about 71.787 thousand km^2. Lake water volume totals 708.8 billion m^3, of which fresh water accounts for 226.1 billion m^3. In the last 40 years, as a result of climate and human activities, the number of lakes has been declining. The water levels and sizes of many inland lakes in northern China are the result of the relationships that exist between precipitation, evaporation, and inflow within the drainage basin. These lakes are rather sensitive indicators of the changes in climate and environment. Data collected on water levels and size of lakes in the region illustrate that many lakes are decreasing in size, and some have dried up (Li and Lin, 1991).

Recession and Thinning of Glaciers. The accumulation and recession of glaciers is mainly the result of long-term changes in climate. At present, most of the glaciers in China are retreating and thinning and the snowline is rising, largely as a result of reduced precipitation and higher temperatures. Data indicate that the glacier areas of northwestern China have been reduced greatly in the last 500 years, and particularly in the last 100 years, when the rate of reduction accelerated significantly (see Table 4).

As shown in Table 4 (Wang, 1992), the glacier area has been reduced by nearly 20%, about 7,000 km^2, in the past 500 years. Further calculation indicates that in the most vigorous period of glacier expansion in the past 500 years, the glacier area reached 35,332 km^2. However, the amount of change differs greatly between glaciers. Smaller glaciers are reduced proportionately faster. Moreover, data analyzed for the past 40 years show that from the late 1950s to the 1980s, about 44.6% of the 23,900 glaciers in northwestern China were retreating, 25.7% were advancing, and 29.7% were either

Mountain Range	Greatest Areal Extent in 500 Years (km²)	Present Area (km²)	Decrease in Past 500 Years (km²)	(%)
Altay	449	293	156	35
Taishan	12,248	9,196	3,052	25
Pamir	2,882	2,207	676	23
Karakorum	6,630	5,925	705	11
Kunlun	9,835	8,735	1,099	11
Qiling	3,288	1,972	1,316	40
TOTAL	**35,332**	**28,328**	**7,004**	**20**

Table 4. Estimated glacier change in the past 500 years in northwest China.

stationary or their classification was unknown. On average, the receding rate was about 10-20 m per year. Recession rates were most rapid (more than 100 m per year) in the Kunlun Ranges.

Persistent Decline of Groundwater Resources. The annual variation of groundwater resources is closely related to precipitation. As a result of persistent drought, groundwater recharge is reduced greatly, resulting in a lower water table. For example, the persistent drought that extended from 1980 to 1984 in northern China resulted in low groundwater levels for 5 consecutive years. Since most of the cities in China are dependent on groundwater, more than 200 cities are now facing water shortages caused by urban population growth and rapid industrial development. Of these cities, 40 are suffering from serious water shortages. In particular, in the last 10 years, groundwater levels have dropped rapidly and many funnels have developed. This situation could be attributed to persistent drought conditions in northern China in recent years (i.e., a dramatic reduction of precipitation in Beijing, Tianjin, Hebei Province, and Shandong Province). Extraction of groundwater has been excessive and replenishment has been insufficient. For example, in the last 10 years, natural water resources have been reduced by 16.2 billion m^3 in the Beijing area. As a consequence, the excessive extraction of groundwater has caused its level to decline at a rate of more than 1 m per year and led to the development of a large funnel of 1,000 km^2. In the central part of the depression area, the groundwater level has declined more than 20 m. In Hebei Province, 23 funnels have been found in recent years, totaling 26,766 km^2, and groundwater levels were lowered by 16.5 m, about 1.4 m per year.

Expansion of Desertified Land

Desertification reflects the lack of harmony between human activities and the environment in arid, semiarid, and some subhumid zones. The degradation of the environment that results is usually attributed to human activities and drought. Desertification is a serious environmental problem in many countries.

China has been seriously affected by desertification. The intensity of desertification can be classified in two categories (Zhu et al., 1989). The first of these is land that has already been desertified. This occupies 52.7% (about 176,000 km^2) of the total area of desertification-prone land. The second category is latent desertification-prone land, or land of potential risk of desertification, covering 47.3% (about 158,000 km^2) of the total area of desertification-prone land. Desertified land areas can be classified into three types according to climatic zone. The first zone is the desertification-prone lands in subhumid zones that make up about 3.9% of the desertified lands. These lands are distributed in narrow and prolonged belt-type areas along river banks, where natural conditions are better than those in arid and semiarid zones. The annual precipitation is usually about 500-600 mm. Desertification in these zones can be reversed if human activities are altered to be more in harmony with the environment. The second zone is the desertified land in the steppe and desert steppe areas of semiarid zones, where the precipitation generally ranges between 250 and 500 mm. Although even a slight disharmony between human activities and the environment will trigger the desertification process in these zones, it can be reversed with proper management, since the moisture and vegetation conditions are slightly better than in the arid desert areas. Reversing the process of desertification is more easily accomplished in eastern China, where precipitation is greater. The third zone is in the arid desert areas located mainly in the western parts of Langshan Mountain, Helan Mountain, and Wushao Ling Mountain; the edges of some large deserts; or around oases. This zone accounts for 30.7% of the desertified land areas in northern China.

It appears that the land area affected by desertification in China will continue to expand because of poor land use practices, the occurrence of drought, and wind erosion. From a comparison of two aerial photographs taken in the 1950s and 1970s, it is estimated that the desertification-prone lands have increased from 137,000 km^2 to 176,000 km^2, an increase of 39,000 km^2 within 25 years, or an average of 1,560 km^2 per year. In addition to the role played by human activities, the apparent aridization trend in northern China is also an important contributor to this expansion of area. Therefore, without appropriate human response, the process of desertification will undoubtedly continue to affect larger areas.

Sea Water Intrusion and Land Surface Sinking

Sea water intrusion is considered to be part of environmental degradation processes and is the result of continued and increasing drought occurrence and the mining of groundwater. In recent years, salt water intrusion has increased in the areas along the coast of the Bohai Sea, such as Shandong, Hebei, and Liaoning Provinces, as well as in the coastal zones of Jiangshu and Zhejiang Provinces. For example, since the severe drought in 1976, salt water intrusion has become increasingly serious along the coast areas of Laizhou Gulf in the Bohai Sea. From September 1987 to August 1988, the area affected by the intrusion reached 435 km^2 in total and the average intrusion rate was 404 m each year, which had a number of adverse consequences. A great deal of farmland was salinized and became nonirrigable. Agricultural output was reduced dramatically, industry suffered great losses, and the quality of drinking water was seriously affected. These consequences can be attributed mainly to the excessive extraction of groundwater because of persistent droughts. Drought conditions during 14 consecutive years (1976-89) were particularly critical. The average annual precipitation during this period was 33% below normal. Drought conditions between 1981 and 1984 were especially critical, averaging only 57% of normal. Although precipitation conditions improved considerably in 1985, four years of severe drought followed. As a result, many rivers, lakes, and reservoirs dried up. The excessive extraction of groundwater that occurred in response to this extended drought resulted in the intrusion of salt water.

Long-term drought and excessive extraction of groundwater have also resulted in the subsidence of the land surface in a dozen large to medium-size cities such as Shanghai and Tianjin. In Tianjin, the maximum subsidence between 1960 to 1983 reached 216 mm per year. In all, a total of about 7,000 km^2 have been affected by subsidence, reaching a maximum of 2.27 m in the center of the subsidence zone. Surface subsidence is a very serious consequence of drought.

MONITORING, WARNING, AND PREDICTION OF DROUGHT

Monitoring and Early Warning

As noted earlier, the onset of drought is gradual and its intensity develops slowly over a prolonged period of time. These features of drought have made monitoring drought conditions a difficult task. In general, two types of monitoring systems exist: surface observation and satellite/remote sensing networks (see Chapters 2 and 3). Using both of these formats, a three-dimensional monitoring system is established that is characterized by combinations of ground-based observation and space-based and mobile observation facilities. These networks are capable of monitoring all the critical elements associated with drought, including precipitation, surface water, soil moisture, crop water requirement, groundwater, irrigation, and drainage. In addition, drought indices can be calculated using data collected from these networks. In drought monitoring, it is essential to establish appropriate and reliable drought indices for different user groups. It is also important to transmit this information in a timely fashion via a suitable telecommunication system. A properly organized delivery system is essential for the monitoring and early warning system to be effective as a disaster mitigation tool. On the basis of the information generated through this system, emergency activities can be better coordinated. This information, used in combination with weather/drought forecasts, can provide decision makers with information needed for disaster management and also provide a scientific rationale for government, agricultural, water conservancy, and other sectors to make short-, medium-, and long-range decisions.

Monitoring by remote sensing techniques is most appropriate for detecting the status of crop growth, soil moisture, evapotranspiration, and precipitation. Using a surface observation station network and remote sensing techniques, the development and spread of drought conditions can be monitored in a routine and cost-effective manner. The Institute of Geography and the Institute of Remote Sensing Application, both of the Chinese Academy of Sciences, have done considerable work in this area. Crop growing status can be monitored through a greenness index derived by linear and nonlinear combination of visible and near-infrared spectra and the ratio between the spectra. Using this greenness index, a model for northern China has been established. An improved thermal-inertia model has also been developed for monitoring soil moisture in areas without vegetative cover. For vegetated areas, the widely used crop water deficit model has been improved, using the principles of crop stress to increase computational accuracy. Substantial progress has also been achieved in the dynamic monitoring of soil moisture and drought in the plains areas of the lower reaches of the Yellow River. The results indicate that surface net radiation, surface temperature, crop evapotranspiration, soil moisture, and other parameters can be estimated by Advanced Very High Resolution Radiometer (AVHRR) graphs of the U.S. National Oceanic and Atmospheric Administration (NOAA) meteorological satellite in combination with surface observational data. The advantage of remote sensing is that the imaging process is fast. When it is used in combination with more conventional methods, surface models, hydrometeorological data, and geographic information system techniques, timely products can be provided to numerous users.

Drought Prediction

In addition to a knowledge of historical studies of drought occurrence and patterns, an understanding of the causes of drought is critical to our ability to predict its occurrence. Drought results from extended periods of below-normal or anomalous precipitation. These precipitation anomalies are the result of large-scale anomalies in atmospheric circulation patterns that become established and persist for months, seasons, or years. Efforts should be made to identify and understand the unique short-term and long-term features of climate that result in drought.

Variations in atmospheric energy are thought to contribute to short-term climate variability. Unlike short-term weather processes, which represent a closed energy system, the energy factors affecting long-term processes are found outside the atmosphere. On this basis, the underlying surface, in particular the ocean surface, which accounts for more than 70% of the earth's surface, is considered to be the source of heat and moisture affecting atmospheric movement. Numerous studies indicate that the thermal anomaly of surface forcing, such as ocean currents, sea surface temperature, sea ice, ice coverage, soil temperature, and humidity, can result in significant changes of atmospheric circulation (Li et al., 1990). Compared with atmospheric circulation, such forcings change slowly, persist for longer periods, and have a longer memory than the atmosphere. The signal from tropical oceans is strongest. ENSO (El Niño/Southern Oscillation) events are the most obvious example of interannual changes in climate. These events may be a key factor in controlling interannual changes in climate on the regional and global scale. Time-lagged teleconnections could be an essential part of future predictions of climate. Recent studies show that ENSO events are associated with drought in northern China, northern Africa, Australia, and India. Greater attention needs to be focused on these relationships in the future.

Charney (1975) has demonstrated that a deterioration of vegetation conditions in grasslands and forests may alter albedo, surface roughness, wind erosion, and soil erosion characteristics of the surface. Through deficiencies of moisture and energy feedbacks, precipitation may be further reduced, with subsequent negative impacts on vegetation conditions (see Figure 4). Numerous studies have shown that much of northern China, in particular the semiarid regions like the Yellow River Valley, Loess Plateau, and the western part of Inner Mongolia, used to be a dense forest. In 850 B.C., the forest coverage in these areas accounted for 53% of the land area (State Science and Technology Committee of the PRC, 1990). In the past several thousand years, particularly the most recent 200 years, forests and grasslands have deteriorated significantly and have contributed to the aridization of northern China. Further studies of this feedback process are needed to fully understand these interrelationships.

DROUGHT MITIGATION TECHNOLOGIES

Experiences from many regions of the world, including China, indicate that many of the impacts associated with drought can be mitigated. This is true even in extremely arid regions where annual precipitation is below 100 mm. For instance, a highly modern production agriculture has been successfully developed in very arid regions such as

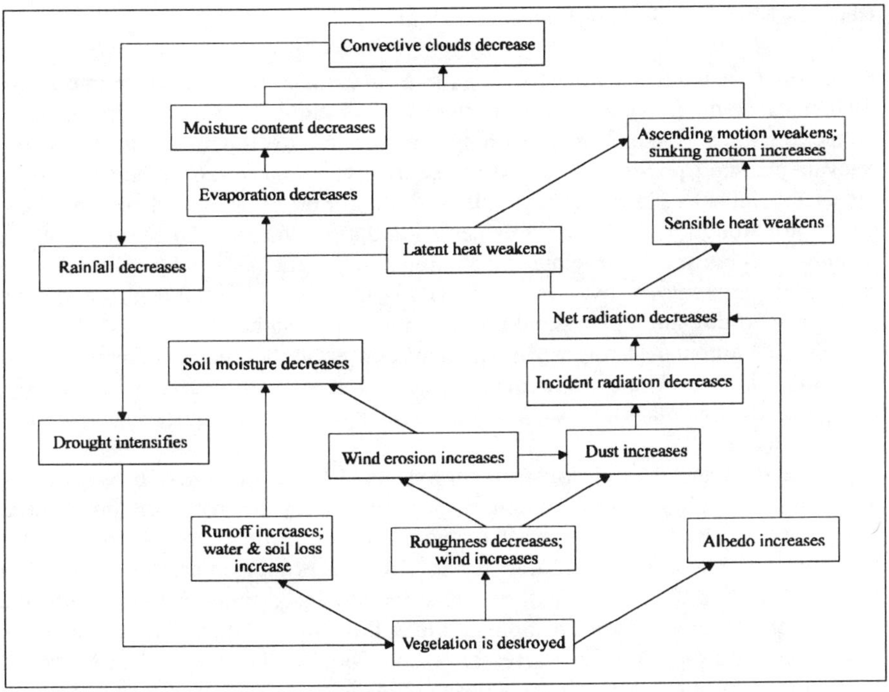

Figure 4. The feedback processes of vegetation and precipitation.

Israel, Saudi Arabia, and the United Arab Emirates. In these regions, grain, vegetable, fruit, and flower production have achieved high outputs, and such products are exported in large quantity. These facts have challenged the conventional concept that it is impossible to develop agriculture in very arid regions. The successful exploitation of agriculture in these regions is of great strategic importance; it is a high point in the human struggle to conquer drought. The American experience in dealing with drought disasters in extensive areas emphasizes that proper policy and planning for drought are an effective way to improve the ability of government to assess and respond to drought (Wilhite, 1991). Actions such as drought management and planning, comprehensive control of drought, water-saving agriculture, dryland farming, and prevention of desertification should be a high priority for China's government.

Drought Management and Planning

Drought management usually refers to actions taken to mitigate the impacts of drought, including assessment and studies of drought information and drought prediction, impacts, monitoring, warning, and response to drought. It also deals with relevant planning, policy making, and necessary institutional arrangements, including the establishment of laws and regulations. Drought planning should be carried out during nondrought periods for implementation during drought. To be effective, the planning process and the development of mitigation programs should involve all of the relevant departments of government. For China, the process of establishing a drought management/mitigation plan should address the following items:

Establishment of Drought Management Centers at the National, Regional, and Provincial Levels. The Centers' major tasks are to design, organize, and implement drought mitigation plans and systematic drought management engineering at national and regional levels.

Establishing a Drought Assessment System. The major tasks of the system are to establish drought indices for monitoring drought, study the spatial-temporal variations of drought and its causes, prepare regional drought climatologies, and perform real-time impact assessment at different stages of drought development. Such a system should be capable of data/information acquisition, storage, display, retrieval, statistical descriptions of drought, and mapping. Since drought is a multidisciplinary issue linking natural and social sciences, research should be interdisciplinary. The establishment of effective drought control and response plans must also integrate these various disciplinary perspectives. In addition to basic meteorological, hydrological, agricultural, and economic data, social and vital statistics data should be included. Studies should also be conducted on the impacts of drought on surface water, groundwater, soil and vegetation, and various social and economic sectors.

Organization and Implementation of Drought Mitigation and Response Programs. The purpose of this step is twofold. It is intended to (1) reduce the potential impacts from drought (i.e., risk management) and (2) reduce personal hardship and impacts during drought and facilitate the recovery process. The establishment of drought mitigation experimental areas to serve as demonstration sites for applying mitigating technology should also be included in this program.

Comprehensive Management of Drought

Combatting drought should be undertaken primarily through protecting ecosystems (Zhang, 1988). This can be accomplished in several ways. First, soil should be protected, since proper water conservation practices can reduce the potential impacts of drought. Second, the vulnerability of each ecoenvironmental system to drought should be established. In other words, research should be conducted on various crops and forage grasses, given the various climatic characteristics of different regions, in order to determine the best or most suitable varieties, both from an environmental and economic point of view. These varieties should be introduced widely to achieve a system integrated between cultivated agriculture, forestry, and animal husbandry. This strategy, coupled with proper management techniques and water conservation projects, can be used to establish a sustainable form of agriculture. In addition, the demand for water in drought areas can be reduced and comprehensive measures can be taken to adjust water diversions within regions. In China and elsewhere, numerous projects of this type have been successful. These experiences, if well documented, can be applied to other regions of the world.

Agricultural Water Conservation Techniques

Water-saving agriculture is suitable for implementation in irrigable but water-deficient areas. This type of agriculture incorporates a wide range of on-farm water conservation techniques into a single scheme (e.g., improved irrigation efficiency, water harvesting, improved cultivation and land management practices, selection of more drought-resistant crops). Water shortages will continue to worsen and will be a significant problem in China for many years. Therefore, improved water conservation practices in both rural and urban settings should be considered as one of China's basic national policies. Agriculture is the major water consumer in China, with an annual consumption of approximately 400 billion m^3, about 80% of the total national consumption. Much of this water is used to irrigate crops. A rather modest reduction in the water used by agriculture, a 10-15% improvement in irrigation efficiency, for example, could save tens of billions of cubic meters of water.

Many research institutes of the Chinese Academy of Sciences and the Chinese Academy of Agricultural Science have conducted a great deal of research on water-saving techniques and demonstrated and promoted these techniques in the Huang-Huai-Hai Plain region. This work has been very successful. It is estimated that if these results were applied throughout the whole region, irrigation efficiency would increase by 20% and grain output would increase by 9.5 billion kg. Expressed in terms of water savings, about 10 billion m^3 would be saved, which is equivalent to the amount provided by 20 large reservoirs or 500,000 wells. These results clearly demonstrate the economic benefits of water conservation.

Dryland Farming

The drought area in northern China, identified by annual precipitation of less than 650 mm, accounts for about one-half of the nation's total area, or a cultivated area of approximately 900 million mu[1]. This represents about 61% of the nation's total cultivated area. About 40% of this area is irrigated; the remainder is nonirrigable. Given the water shortage problems of China and the nation's ongoing problems with drought, it is critical that agricultural water conservation measures be applied to irrigated lands and that scientific dryland farming be widely used on nonirrigated lands.

The primary purpose of dryland farming is to increase the efficiency of precipitation by improving infiltration and reducing evaporation from the soil surface. Most parts of China are dominated by a monsoon climate that is characterized by the uneven distribution of annual precipitation. Therefore, the water supply in the dry season or during drought periods depends mainly on the precipitation accumulated in the soil root zone during the rainy season. The soil's water-holding capacity is immense. According to figures calculated by Lou (1989), the water-holding capacity of 2 m may be as high as 550-600 mm (i.e., one mu of farmland may store 360-400 m^3 of water). In other words, about 9 million mu of cultivated land in northern China may store 330-540 billion m^3 water, roughly equivalent to the total surface water resource, 510 billion m^3, in northern China. Therefore, soil is considered to be a reservoir, and many measures are applied to increase the volume of water stored in the soil.

In recent years, the research institutes that are part of the Chinese Academy of Sciences and the Chinese Academy of Agricultural Sciences have undertaken a number of systematic studies on dryland farming. Experiments for fixed point observation at multiple sites have been conducted on the moisture dynamics of farmland, moisture balance, and agricultural potential productivity for many years. These studies have analyzed the spatial-temporal variation of moisture in dryland soil, periods of moisture deficiency, and characteristics of drought and soil moisture deficiency. In addition, the evapotranspiration rates of the major crops have been determined, along with water use efficiency, crop coefficients, and the value of additional precipitation. Through these studies, a number of approaches for increasing water efficiency have been suggested, such as adjusting agricultural structure, fertilizing soil, optimizing crop patterns, and using chemical preparations. In addition, a number of techniques for adjusting and controlling moisture on various farmlands have also been studied and have achieved significant results for saving water and increasing productivity. The overall results show that a 60% potential increase of agricultural output is possible in the arid area of northern China.

Recovery and Control Measures of Land Desertification

Based on experiences in working with desertified land areas (Zhu et al., 1989), the rehabilitation process should focus on two objectives—ecological improvement and

[1] *Mu* is the Chinese unit of area. One mu = 0.0667 ha.

economic benefits. These objectives should strive to develop a more appropriate (sustainable) system of land use. Decisions regarding the development and use of desertification-prone land areas should be made with careful consideration given to protection and rehabilitation. Two models for the rehabilitation of desertified lands are the readjustment model, to be used in cases of latent and ongoing desertification, and the rehabilitating model, for application to the most severely desertified land areas, where the focus is on reestablishing the ecosystem through the use of new energy sources to help mitigate the effects of vegetation destruction caused by the use of wood for cooking and heating (Zhu, 1990).

Strategies to mitigate desertification should be considered in a systematic way. These strategies represent complex engineering problems that require an integrated, interdisciplinary approach to be successful. It is critical that a master plan be developed for the rehabilitation of ecosystems in arid and semiarid zones. Given the importance of economic development in China, afforestation and animal raising should be emphasized, along with the creation of a diversified economy. Simultaneously, population growth should be controlled in order to reduce the pressure on the land. Energy efficiency in desertified land areas should be improved to put an end to the undue collection of fuelwood and the destruction of vegetation.

Among specific measures to combat desertification in marginal dry farming areas, particularly in semiarid zones, it is imperative that China (1) adjust current dryland farming systems, (2) adopt natural preservation (i.e., stop inappropriate economic activities so that the desertification process gradually will be terminated and the semiarid ecoenvironment will regain its potential to reverse the process naturally), (3) expand afforestation and land areas used for the production of forages, (4) more intensively manage the best arable lands using appropriate soil and water management practices, and (5) combine shelterbelts and tree networks to include planting sand-holding species on dunes. Tree patches should also be established in the interdune areas to reduce wind velocity and stabilize mobile dunes. In the rangeland areas, rational carrying capacity should be determined and proper rotation grazing practiced. Grassland and forage farms should be established as a supplementary source of natural grassland. In addition, domestic wells should be intensively managed and the density of grazing areas and the building of paved roads on pasture lands more effectively controlled. Natural conservation (eliminating human interference) is an effective way to limit the spread of desertification; it has been widely adopted in the semiarid zone. Even in areas with an ongoing desertification problem, natural conservation plays an important role in the rehabilitation process. For example, in the land adjacent to Zhaohai Miao in Horgin Zuoyi Huoqi County in the eastern part of Horgin Sandy Land, this strategy was applied to desertified land for three years. Because of the potential for

Year Coverage and Biomass Community	Without Conservation (Comparison)		1 Year Conservation		2 Years Conservation		3 Years Conservation	
	Veg. Cover (%)	Bio-mass (g/m²)	Veg. Cover (%)	Bio-mass (g/m²)	Veg. Cover (%)	Bio-mass (g/m²)	Veg. Cover (%)	Bio-mass (g/m²)
Artemisia frigida	30	130	45	228	50	255	65	271
Artemisia halodendron	30	483	40	642	45	755	45	733
Weed spp.	40	219	55	310	60	321	70	386
Legmus chinensis	35	183	40	304	60	392	65	364

Table 5. Change of vegetative coverage and biomass since natural conservation in Zhaohai Miao District of Horqin Zuoyi Houqi County (% and g/m²).

naturally reversing the desertification process in this region, vegetation was gradually regenerated without the application of other supplementary measures, and biomass increased, as shown in Table 5 (Zhu et. al, 1989).

Table 6 shows environmental changes that have occurred since the rehabilitation of desertified land in the semiarid zone (Zhu et al., 1989). In the arid zone, river basins must be considered as an ecological unit from a planning perspective. Water resources should be reasonably distributed from the upper to lower reaches of the river. In an oasis setting, protective networks for farmland should be established inside the oasis, with sandbreaks established at the oasis periphery. Sand-stabilizing belts should be established on the land surrounding the oasis to form a complex controlling system. Table 7 shows the changes of surface features around an oasis before transformation and after intervention (Zhu et al., 1989). The changes in surface features indicate a decrease in the amount of severely desertified land. This table also indicates that some desertified lands have been transformed into woodlots and orchards. Thus the productive potential of deteriorated lands has been restored and the desertification process is being reversed.

CONCLUSIONS AND RECOMMENDATIONS

Compared to other natural disasters, drought is considered the greatest disaster in China. This chapter described the regional characteristics of drought occurrence in

Various Landforms & Their Percentage in the Total Area:	Mobile Sands and the Affected Area	Slightly Eroded Fields	Shifting Sands Sparely Distributed	Slightly Desertified Fields
Before the Plantation	44	25	9	22
Since the Plantation	8	4	6	82

Table 6. Environmental changes since the renovation of desertified land in Sanjiazi District, Zhangwu County.

China and the associated impacts of drought on agriculture and the natural environment. In order to reduce vulnerability to droughts in the future, assessment and response efforts must be improved. We propose that drought management centers be established at the national, regional, and provincial levels. The primary function of these centers should be to design, organize, and implement integrated management plans. The activities of these centers would range from the improvement of monitoring techniques and the development of suitable indices to the development of comprehensive mitigation and response programs. Given the pronounced regional differences in drought frequency and impacts, the character of these management plans must be adjusted to address provincial differences, but they also must be well coordinated at the regional and national level.

Landform Patterns	Percentage of the Areas Occupied: Before Transformation	After Intervention
Most Severe Land Desertification	17.8	0.4
Severe Land Desertification (Mobile Dunes)	54.6	9.4
Reversed Desertification (Orchards and Woodlots)	9.0	52.4
Farmland without Desertification Impacts	8.9	37.8
Wastelands	9.7	

Table 7. Changes in surface features of land desertification at Linze Desert Experiment Station of the Institute of Desert Research, Academia Sinica, before transformation and after intervention.

ABOUT THE AUTHORS

Kerang Li is a professor of the Geographic Institute of the Chinese Academy of Sciences, vice-director of the Study Commission for Marine Geography of the Geographical Society of China, a member of the Commission for National Climatology Research of China, and Rapporteur on Drought of the Commission for Climatology of WMO. He graduated from the Department of Geophysics, Beijing University, in 1960 with a specialty in meteorology. Since graduation, his areas of study have included marine climate, sea-air interaction, climatic change and its impact, and drought and floods in China. His present work centers on drought assessment and drought prediction in China.

Xianchao Lin graduated from the Meteorology Department of Zhongshan University in 1982 and gained a master's degree in climatology from the Chinese Academy of Sciences in 1985. He works as a research associate in the Institute of Geography of the Chinese Academy of Sciences. His main research fields are physical-dynamic climatology, monsoon, climatic change, and drought.

REFERENCES

Charney, J. 1975. Dynamics of deserts and drought in the Sahel. *Quarterly Journal of the Meteorological Society* 101:193-202.

Chen Yiqiong. 1986. Drought and impacts [in Chinese]. *Journal of Catastrophology*, 1(1):15-22.

Feng Peizhi; Li Cuijin; and Li Xiaoqun. 1985. *Main Meteorological Disasters in China* [in Chinese]. China Meteorological Press, Beijing.

Li Kerang; and Lin Xianchao. 1991. Drought, China's greatest natural disaster. *Drought Network News* 3(3):3-6.

Li Kerang; and Lin Xianchao. 1992. Drought, China's greatest natural disaster. *Drought Network News* 4(1):7-10.

Li Kerang; Xu Shuying; and Guo Qiyun. 1990. *Drought and Flood in North China Plain* [in Chinese]. China Science Press, Beijing.

Lou Chenghou, ed. 1989. *Agricultural Modernization of Dryland in the Northern of China* [in Chinese]. China Meteorological Press, Beijing.

Ministry of Water Conservancy, Hydrological Department. 1987. *Water Resources Assessment for China* [in Chinese]. Water Conservancy and Electric Power Press, Beijing.

Shen Zhenrong. 1982. Time-space distribution of drought recent 30-year in China [in Chinese]. *Natural Resource* 4:9-18.

Song Lianchun. 1991. A study on the summer drought/flood and its impact in China during 1951-1990 (in Chinese). *Journal of Nanjing University*, pp. 208-15. Special issue on the causes and countermeasure of natural disaster.

State Science and Technology Commission of the PRC. 1990. *Climate*. China Meteorological Press, Beijing.

Wang Zhongtai. 1992. Glacier change and its impact since the little ice age and their future trends in Northwest China [in Chinese]. *Studies on Climatic and Sea Level Changes in China, II*, pp. 121-22.

Wilhite, Donald A. 1991. Drought planning: A process for state government. *Water Resources Bulletin* 27(1):29-38.

Zhang Jiacheng. 1988. *Climate and Mankind* (in Chinese). Henan Scientific and Technical Press.

Zhang Xiangong. 1981. Analysis of drought index in recent 500-year in the east of China (in Chinese). *Collected Works of Climatic Change Symposium.*

Zhu Zhenda; Liu Shu; and Di Xinmin. 1989. *Desertification and Rehabilitation in China*. China Science Press, Beijing.

Zhu Zhenda. 1990. Desertification and its control in Northern China. *Recent Development of Geographical Science in China*, pp. 67-75. China Science Press, Beijing.

Index

A

C

D

E

F

G

H

I

L

M

N